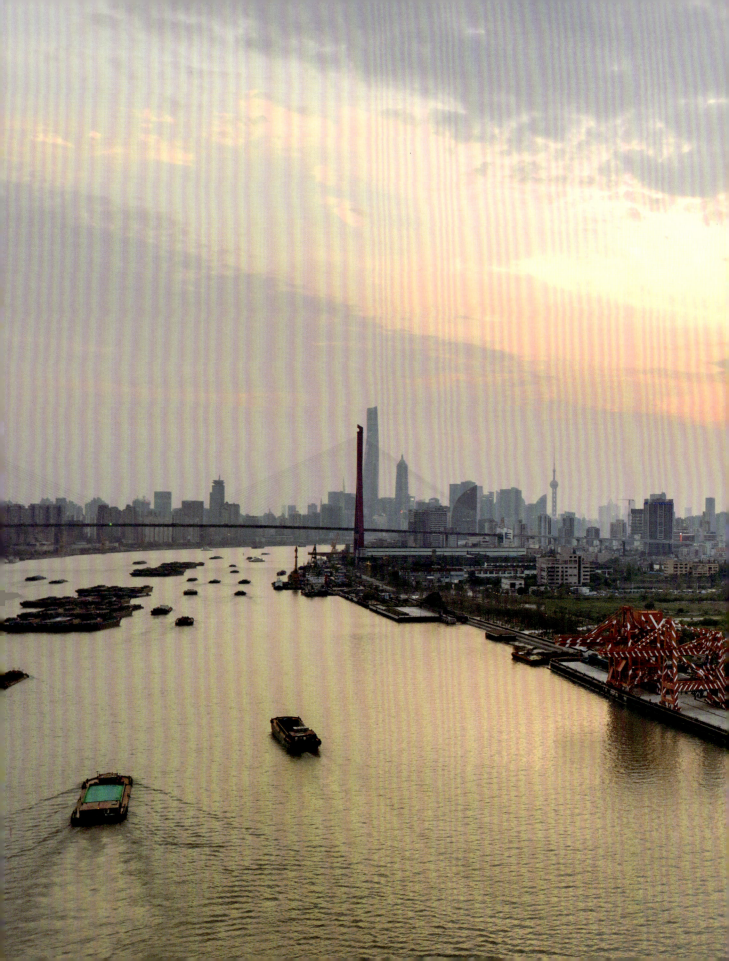

相 遇

2019
上海城市空间艺术季
主展览

《相遇》编委会 编
Edited by *Encounter* Editorial Board

Encounter

2019 SHANGHAI URBAN SPACE
ART SEASON MAIN EXHIBITION

东华大学出版社·上海
Donghua University Press Shanghai

主办
上海市规划和自然资源局
上海市文化和旅游局
上海市杨浦区人民政府

HOSTS
Shanghai Urban Planning and Natural Resources Bureau
Shanghai Municipal Administration of Culture and Tourism
Yangpu District People's Government

学术委员会

为确保空间艺术季的专业性、学术性和国际性，特成立空间艺术季学术委员会，为空间艺术季各项核心工作提供专业指导和评审意见。委员会体现国际性和专业性，委员均为国际和国内专业造诣高、工作经验丰富且有影响力的专家，并保持国际专家占一定的比例。委员会为多专业的跨界合作，充分体现空间艺术季是专业性、公众性相结合的大型城市公共活动。

职 责
学术委员会职责包括召开各阶段学委会会议，审议有关工作事项，进行策展人和策展方案评审等；活动期间参与论坛和研讨会，进行主旨演讲和学术交流等；出席空间艺术季期间各重要节点活动等。

主 任
本届学术委员会主任由中国科学院院士郑时龄担任。

委 员
学术委员会委员由城市规划、建筑、景观、公共艺术、传播、评论、策展、社会学、媒体等领域专家组成，人数约 30 人。学术委员会委员可根据每届的具体情况做局部调整。

ACADEMIC COMMITTEE

To guarantee that SUSAS is a professional, academic and international event, the academic committee is established to offer professional guides and evaluative opinions on various core tasks of the art season. Committee members, domestic and abroad, feature expertise, experiences and influence, and a considerable of them are internationally renowned figures. The committee is comprised of members from multiple disciplines, fully demonstrating SUSAS as a major urban public event that is both professional and open.

RESPONSIBILITY
The responsibilities of academic committee include: organizing academic committee sessions at each stage where progress, curators and plans are discussed and evaluated; attending forums and seminars during the art season, giving keynote speeches and participating in academic dialogues; attending various key activities during the art season, etc.

PRESIDENT
Zheng Shiling, fellow of the Chinese Academy of Sciences.

MEMBERS
The members of academic committee cover about 30 professionals in urban planning, architecture, landscape, public art, communication studies, art criticism, curation, sociology and press. The list is subject to moderate adjustment according to the requirements of each year.

学术委员会主任	PRESIDENT
郑时龄	Zheng Shiling

学术委员会委员	COMMISSIONERS
伍江	Wu Jiang
丁乙	Ding Yi
李磊	Li Lei
汪大伟	Wang Dawei
俞斯佳	Yu Sijia
赵宝静	Zhao Baojing
方晓风	Fang Xiaofeng
管怀宾	Guan Huaibin
顾建军	Gu Jianjun
顾骏	Gu Jun
凯瑟琳·摩尔	Kathryn Moore
李龙雨	Yongwoo Lee
陆蓉之	Lu Rongzhi
李翔宁	Li Xiangning
娄永琪	Lou Yongqi
马兴文	Ma Xingwen
秦畅	Qin Chang
孙玮	Sun Wei
吴海鹰	Wu Haiying
王建国	Wang Jianguo
吴蔚	Wu Wei
张宇星	Zhang Yuxing
朱国荣	Zhu Guorong

目 录

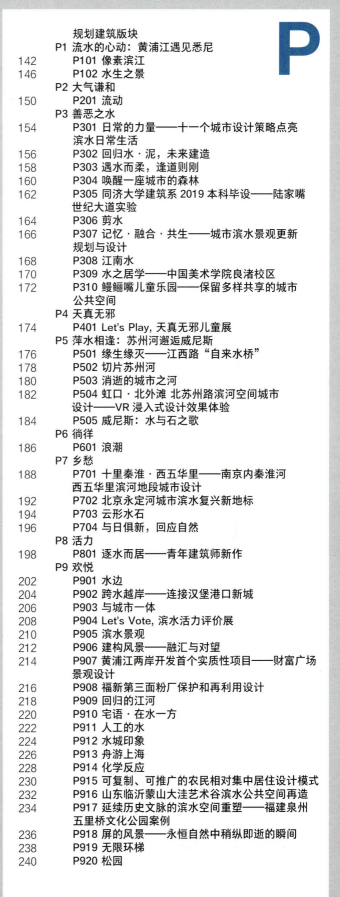

Contents

A

P

关于 2019 上海城市空间艺术季
Shanghai Urban Space Art Season 2019

上海城市空间艺术季（以下简称"空间艺术季"），继 2015 年、2017 年成功举办两届以来，2019 空间艺术季由上海市规划和自然资源局、上海市文化和旅游局、杨浦区人民政府共同举办，继续以上海的城市更新发展阶段为背景，探讨城市空间话题。

空间艺术季继承发扬"城市，让生活更美好"的世博精神，以"文化兴市，艺术建城"为理念，以"城市艺术化，艺术生活化"为目标，通过"艺术植入空间""展览与实践"相结合的方式，将公共艺术和城市更新的实践成果引入展览，将举办空间艺术季取得的收获应用于城市更新的实践中，实现活动每举办一届，文化热点就传播一次，国内外大师作品就留下一批，城市公共空间就美化一片的目的，打造"永不落幕的世博会"。"城市空间"既是上海城市空间艺术季的展场，也是上海城市空间艺术季的重要展品，通过活动的举办让市民感受城市空间品质提升的艺术魅力，这是空间艺术季有别于其他文化艺术活动的重要特点。

2019 上海城市空间艺术季立足"一江一河"公共空间提升战略，通过艺术植入空间的活动，邀请人们亲身感受滨江贯通这一公共空间作品；通过展览与实践相结合的方式，搭建一个对话的平台，聚焦"滨水空间为人类带来美好生活"这一世界性话题。通过两个多月的主展览、实践案例展、联合展和各类公众活动，结合"百万市民看上海"等文旅融合策划，向全世界介绍了上海近年来遍布全市的各类滨水空间贯通、品质提升的建设成就及未来愿景。

主题——"相遇"

主展主题由总策展人提案确定为"相遇"，意喻本届艺术季所呈现的人与人的相遇、水与岸的相遇、艺术与城市空间的相遇、历史与未来的相遇，将激发更多美好生活、美好情感的相遇。

主展览

2019 上海城市空间艺术季主展览以原上海船厂旧址地区（包

Following two successful sessions in 2015 and 2017, the Shanghai Urban Space Art Season (SUSAS) 2019 is co-organized by the Shanghai Urban Planning and Natural Resources Bureau, and the Yangpu District People's Government. SUSAS 2019 continues to discuss topics related to urban spaces based on the current urban renewal in Shanghai.

As a continuation of the EXPO 2010 spirit "Better City-Better Life", SUSAS, guided by the idea of its concept "Culture Enriches City, Art Enlightens City", aims for "a city of art" and "art of life" through "implanting art into spaces" and "exhibition + site projects". In this way, practices of public art and urban renewal are introduced into exhibitions while the accomplishments of SUSAS may be applied to urban renewal at work. Each SUSAS is intended, as an "EXPO that never closes", to make a hot cultural topic, preserve masterpieces created by international artists and beautify a section of the city. Urban Spaces are more than where exhibitions are held, but are essential exhibited items of SUSAS, allowing citizens to experience how the spaces are enhanced aesthetically. This is a key feature that sets SUSAS apart from other cultural events.

SUSAS 2019 is based on the strategy of constructing public spaces along the Huangpu River and the Suzhou Creek. It invites people to Huangpu River Connection as a public space by implanting art into spaces and builds a platform for dialogue on the global issue "waterfront space brings a better life of mankind" by incorporating exhibitions with practices. In about two months, via main exhibition, site projects, joint exhibitions, and various public activities, in collaboration with cultural tourism programs such as "Millions of Citizens Observe Shanghai" etc., SUSAS shows the world how Shanghai has connected and enhanced the city's myriad of waterfront spaces as well as its future vision.

Theme of SUSAS 2019: Encounter

The theme of the main exhibition, as proposed by the chief curator, is Encounter, representing the encounter between people, water and bank, art and urban space, history and future, which will further create a better life and inspire more emotional exchanges.

Main Exhibition

The venue of the SUSAS 2019 Main Exhibition is the old site of the Shanghai Shipyard, including the docks and the Maoma (linen and wool) warehouse, while the outdoor public artworks are displayed along the 5.5-kilometer waterfront public space from Qinhuangdao Road to Dinghai Road. The main exhibition includes sections of public artworks, planning and architecture, and urban space art.

The chief curator of the main exhibition is Fram Kitagawa, an internationally renowned artist and curator who has long been dedicated to boosting the regional economy with art. The Urban Space Art section, headed by Fram Kitagawa and assisted by Ms. Wang Bin, director of Xinzhifeng Art Institute, is an artistic anchor along the 5.5km of Yangpu waterfront space, with its 19 permanent installments as on-site

括船坞和毛麻仓库）作为本届艺术季的主展展场和展馆，以杨浦区滨江南段5.5公里滨水公共空间(从秦皇岛路至定海路）作为户外公共艺术作品的延伸展场。主展览通过公共艺术作品、规划建筑版块和空间艺术版块，演绎滨水空间的话题。

主展总策展人由长期致力于艺术振兴区域经济实践的国际知名艺术家、策展人北川富朗先生担任。城市空间艺术版块，以北川富朗先生为主，欣稚峰艺术机构负责人汪斌女士配合，邀请世界各国艺术家，结合杨浦滨江的历史底蕴、空间特点和未来展望，在地创作 19 件公共艺术作品留存在 5.5 公里滨江空间中，成为滨江空间的艺术焦点。在船坞和毛麻仓库展区，结合"相遇"主题和建筑空间特点举办艺术展。与此同时，本届空间艺术季还提供 5 个公共艺术品点位，向世界募集优秀作品，募集遴选出的优秀作品的其中一件作品也制作落成，留存于滨江空间内。

规划建筑版块策展人由上海交通大学设计学院院长阮昕教授担任。该版块生动演绎国内外滨水空间的规划建设理念，综合呈现上海"一江一河"建设的成就和未来展望，丰富本届空间艺术季的知识性和趣味性，激发人们更多的理性思考，引导形成对滨水空间发展目标的社会共识。

本届空间艺术季由时任同济大学建筑系副主任章明担任总建筑师，对艺术季重要作品船坞在展期内的再利用进行整体建筑设计，并会同参与设计和建设杨浦滨江公共空间的各建筑师团队，包括原作设计工作室、同济大学建筑设计研究院（集团）有限公司都境建筑设计院、致正建筑工作室、刘宇扬建筑事务所、大舍建筑设计事务所、上海大观景观设计有限公司、优德达城市设计咨询有限公司等团队共同完成展场设计，打造艺术季的整体形象。

实践案例展

实践案例展以实地呈现的方式，为市民介绍近年来本市优秀滨水空间建设案例，在地举办市民活动。本届共涉及 7 个区 13 个案例展。其中，黄浦江沿线的案例展有浦东新区实践案例展、徐汇区西岸实践案例展、闵行区浦江第一湾公园实践案例展；苏州河沿线的有虹口区滨水空间实践案例展、普陀区 M50 实践案例展、长宁区苏州河实践案例展；骨干河道沿岸的有静安区彭越浦河岸景观改造实践案例展；古城水系沿岸的有嘉定环城河步道城市更新实践案例展；新城水系沿岸的有青浦环城水系公园空间艺术实践案例展、松江区华庭湖实践案例展、奉贤新城实践案例展；水乡村庄的有金山区漕泾镇水库村实践案例展；生态湿地的有崇明东滩湿地公园实践案例展。

本届艺术季还首次探索了文旅融合模式，结合"百万市民看上海"等文旅融合品牌活动，带领市民共同领略这些滨水空间的魅力。

productions of international artists with respect to local history, character, and vision. In the dock and the Maoma warehouse, exhibitions are held in conjunction with the theme of "Encounter" and the characteristics of the architectural space. Meanwhile, to motivate artists interested in urban public art to participate, SUSAS 2019 has initiated an international call-for-work for five slots of public art, and one of the outstanding works selected from the collection is also completed and now becomes a permanent installment at the Yangpu Waterfront.

The curator of the Planning and Architecture section is Ruan Xing, Dean of School of Design, Shanghai Jiaotong University. This section vividly demonstrates the concepts of waterfront planning at home and abroad as well as the achievements and prospects of construction projects along the Huangpu River and Suzhou Creek. Being an intellectually interesting exhibition, this section intends to encourage people to think more rationally and form a consensus on the aims of waterfront development.

Zhang Ming, the Vice Head of the Department of Architecture, College of Architecture and Urban Planning, Tongji University, serves as the chief architect of SUSAS 2019. He is responsible for the overall architectural design of the reutilized Dock which is an essential exhibited item and the venue design with other architects involved in designing or constructing the public spaces of Yangpu Waterfront, establishing the overall image of SUSAS 2019. The team of architects includes Original Design Studio, Dujing Architectural Design Institute of Tongji Architecture Design (Group) Co., Ltd., Atelier Z+, Atelier Liu Yuyang Architects, Atelier Deshaus, Da Landscape and UrbanDATA Design Consulting Co., Ltd..

Site Projects

The site projects, presented in the field, introduce several recent exemplar waterfront projects, involving citywide 13 projects in 7 districts. Among them, along the Huangpu River are site projects at Pudong New Area, Shanghai West Bund of Xuhui District, and Pujiang First Bay Park of Minhang District; along the Suzhou Creek are the waterfront space site project at Hongkou District, the M50 site project of Putuo District and the site project of Suzhou Creek waterfront in Changning District; along the backbone river is the Pengyuepu riverbank landscape reconstruction practice case exhibition; along the ancient city water system is Jiading moat promenade urban renewal site project; along the new city water system are Qingpu round-city water park space art practice, Songjiang Huating Lake site project, and Fengxian New City site project. In addition to these, there are site projects in water village and ecological wetland, respectively the site project of Shuiku Village of Jinshan District and Dongtan Wetland Park site project at Chongming District.

The 2019 art season explores for the first time the integration mode of cultural tourism, leading the public to experience these charming spaces in collaboration with "Millions of Citizens Observe Shanghai" and other programs involving both cultural and tourist elements.

联合展

联合展邀请社会各界自发组织与空间、艺术、城市更新等相关的公共艺术展和群众文化活动，在空间艺术季举办期间同步开展，共同倡导城市空间艺术美。

三届空间艺术季中，联合展共推出了 28 个展览项目，参与主体约 20 个，主题丰富多样。联合展践行了空间艺术季跨出行业领域，跨界联合的理念，拓展了空间艺术季朋友圈范围，提升了空间艺术季的整体社会活力。

2019 空间艺术季联合展共 13 个，按展览展示内容可分为城市更新和公共艺术两大类型。其包括"相遇·贵州路""陆家嘴滨江金融城公共景观装置展""花开上海"3 个城市更新类，以及"第六届全国大学生公共视觉优秀作品双年展""步履不停：1995-2019 年中国当代艺术的城市叙事""隽永墨韵大华银行水墨艺术展""合成空间＆感官""盲点艺术展""未来感官——重塑城市空间的光与形灯光展""生境花园：给城市野生动植物建个家""符号上海·相遇水岸""2019 多瑙河对话艺术节梦之声——丝路上的中国当代艺术""'渡·爱'2019 外滩艺术计划"，共 10 个公共艺术类展览项目。本届联合展首次采用了社会征集展览项目的方式，使联合展真正地成为了一个开放的平台。

SUSAS 学院

公众活动（SUSAS 学院）通过系列论坛、讲座和文艺演出等公众活动，以空间艺术季的主题和空间为载体，为专业人士提供交流平台，为市民参与艺术季的创作提供契机。2019 空间艺术季的公众活动（SUSAS 学院）版块围绕"滨水空间给人类带来美好生活"的主题方向，通过学术交流、市民宣传、儿童教育及其他活动策划并组织活动，让不同专业领域、不同年龄层次的人在主展场空间"相遇"，讨论"滨水空间"的更新，体验"城市艺术"的魅力，感知"杨浦滨江展区"的历史，展开对"未来生活"的畅想。

2019 空间艺术季在面向公众展览的 80 天里，共组织 144 场活动。其中，学术交流活动 20 场，如大师讲坛、同济设计周系列讲座、设计教育会议、H+A 华建筑学术年会等；市民宣传活动 10 场，如公共艺术作品公众导览、滨江空间景观公众导览、社区规划师沙龙、杨浦滨江摄影大赛与优秀作品展、城市空间行走体验、艺术护照打卡等；儿童教育活动 78 场，如小小导赏员、小小观察员、小小建筑师、小小规划师、儿童创意工作坊、乐高积木搭建等；其他类型活动 35 场，如绿之丘临展、大学生创意市集、种子魔方活动、滨江定向赛等。

大师讲坛为学术交流提供了良好的专业平台；快闪活动，为公众提供认知城市空间与公共艺术的趣味方式；儿童活动，为不同年龄段的少年儿童提供了美育乐土；企业驻场，为高校、新生代专业机构提供了展示舞台。

Joint Exhibitions

The joint exhibitions invite all sectors of the society to organize public art exhibitions and cultural events related to space, art, and urban renewal, which are carried out during the art season to jointly advocate the artistic beauty of urban space.

During the three sessions of SUSAS, the joint exhibitions have launched a total of 28 exhibition projects, with about 20 participants and a rich variety of themes. The joint exhibitions put into practice the concept of transcending industry and cross-field cooperation, expanding the scope of industry interaction and enhancing the overall social vitality of SUSAS.

SUSAS 2019 has 13 joint exhibitions, which can be divided into two types: urban renewal and public art according to the content. There are three of the urban renewal, including "Encounter, Guizhou Road", "Public art for Lujianzui Harbor City", "Blooming Shanghai-Community Garden Plan 2040"; then ten public art exhibitions are "The 6th national art institutes and academic biennale for outstanding public visual artwork of graduates and undergraduates", "A turning moment: city narrative of Chinese contemporary art from 1995 to 2019", "UOB art in ink exhibition", "Synthetic spaces & Sensorium", "Blind Spot", "LIGHTENING the living future", "Habitat Garden: let's give wildlife a home in our city", "Iconic Shanghai Encounter · Waterfront", "2019 Danube dialogues dream sound - Chinese contemporary art on the Silk Road", "The bund art project 'Ferry Love' 2019". This year's SUSAS adopts an open call for joint exhibitions for the first time, making the joint exhibitions a true platform for all.

SUSAS College

The public activities under the title of SUSAS College whose venues are supplied by SUSAS, including serial forums, lectures, and shows, follow the theme of SUSAS 2019. They provide professionals with a platform to exchange ideas and the citizens with opportunities to play a part in the creative activities of SUSAS. SUSAS College plans and organizes activities around the theme of "Waterfront Space Brings Better Life to Humanity" through the academic exchange, citizen publicity, child education, and other activities, allowing people of different professions and ages to encounter, discuss the renewal of waterfront space, experience the charm of urban art, perceive the history of the Yangpu Riverfront Exhibition Area, and imagine the future life.

A total of 144 events have been organized during the 80 days of the SUSAS 2019. Among them, there are 20 academic exchange activities, such as masters' lecture, series of lectures of Tongji Design Week, design education conferences, the academic annual conference of H+A Hua Architecture, etc.; 10 citizen publicity activities, including public artwork guide, landscape guide of waterfront space, community planner salon, Yangpu Waterfront photography competition and outstanding works exhibition, Walking in Urban Space, Design Your Art Passport, etc.; 78 child education activities, such as experience programs to play the role of tour guide, observer, architect or planner, Children Creative Workshop, Build with Lego, etc.; and 35 other activities, including the Green Hill temporary exhibition, Creative Market for College Students, Seed Cube Game, Riverside Orienteering, etc..

The masters' lecture provides a good professional platform for academic exchange; the flash mob gives the public a fun way to perceive urban space and enjoy public art; the children's activities offer a great opportunity for aesthetic education for children of different ages; and the enterprises are on-site during the event, enabling the universities and new-generation professional institutions to showcase themselves.

In partnership with China Shanghai International Arts Festival, Shanghai Citizens Art Festival, and Shanghai Tourism Festival, and with the quadruple aims of "Service in Shanghai, Made in Shanghai, Shopping in Shanghai, Culture of Shanghai", SUSAS offers its venues for diversified art-educational activities for the citizens, affording lovely life with lovely spaces.

同时，空间艺术季还与上海国际艺术节、市民文化节、上海旅游节合作，融合"上海服务、上海制造、上海购物、上海文化"四大品牌建设目标，以空间艺术季的展场为载体，推出丰富多彩的市民美育活动，以美好的空间承载美好的生活。

序 1
空间赋能 艺术建城——以空间艺术季推动人民城市建设的上海城市更新实践

Preface 1
Energized by Space, Built by Art—Shanghai Urban Renewal Practice Promoting the Construction of People's City with SUSAS

徐毅松
上海市规划和自然资源局局长

Xu Yisong
Director of Shanghai Urban Planning and Natural Resources Bureau

2019 年 11 月 2 日，习近平总书记深入上海杨浦滨江等地，就贯彻落实党的十九届四中全会精神、城市公共空间规划建设和社会治理等方面进行调研，特别在视察杨浦滨江时做出"文化是城市的灵魂""人民城市人民建，人民城市为人民"等重要指示。以杨浦滨江地区空间改造为代表的上海城市更新实践，在规划理念上注重空间赋能，充分挖掘和发挥空间自然要素附加的生态价值、人文价值和经济价值；在工作实践中注重以点带面，以空间艺术季这类"文化事件"有效激活空间更新的持续活力；在实施机制上注重众筹共治，通过统筹政府、市民和市场等多主体共同协商，不断提升市民在人民城市建设过程中的参与感，不断增强市民对城市空间改善的获得感和幸福感，不断提高规划资源在现代化国际大都市空间治理领域的能力水平。

以"空间艺术季"为载体向滨江空间赋能

新时代的规划资源管理，始终以国土空间源头治理推动城市高质量发展和引导人民高品质生活为己任。正是在这样的理念引导下，上海在城市更新工作中尤其注重空间之于城市生态、地区经济及社会人文的载体作用，并开展了一系列内涵丰富的城市更新实践活动。上海城市空间艺术季（英文简称"SUSAS"）就是其中之一。

艺术季继承发扬"城市，让生活更美好"的世博精神，虽然以展览为形式，但通过空间艺术布展与城市有机更新实践的相互推动，赋予并展示了城市空间更多的内涵和价值。空间艺术季每两年举办一次，活动内容由空间场馆改造主展、实践案例展、地区联合展、公众活动（SUSAS学院）等版块组成，旨在实现"举办一届展览活动、传播一次文化热点、留下一批大师作品、美化一片城市空间"的目的，打造"永不落幕的世博会"。目前连续举办的三届展览活动均选址于黄浦江沿岸地区，伴随"人民之江"建设和滨江地区更新进程，对滨江空间结构优化和品质提升发挥了积极作用。

On November 2, 2019, General Secretary Xi Jinping went to Yangpu Riverfront in Shanghai to research the implementation of the spirit of the Fourth Plenary Session of the 19th CPC Central Committee, urban public space planning and construction, social governance, etc.. During his inspection of Yangpu Riverfront, he made a few important instructions, such as "Culture is the soul of the city", "People's city should be built by the people and for the people". Represented by the space reconstruction of Yangpu waterfront area, Shanghai's urban renewal practice focuses on energizing the city with space in its planning concept, giving full play to the ecological, humanistic and economic values of the natural space; in the work practice, it effectively activates the continuous vitality of space renewal with such "cultural events" as SUSAS; as for its implementation mechanism, by coordinating the joint power of multiple subjects such as the government, citizens and the market, it continuously enhances the participation of citizens in the process of building the people's city, increases people's sense of acquisition and happiness for the improvement of urban space, and improves resource planning in the field of space governance of a modern international metropolis.

Energize the Riverfront Space via SUSAS

In the new era of planning and resource management, it is always our mission to promote high-quality urban development and provide people with high quality of life with the source management of national land and space. It is under such a philosophy that Shanghai pays special attention to the importance of space as a carrier of urban ecology, regional economy, and social humanity in the process of urban renewal, and has carried out a series of rich urban renewal practices, one of which is the Shanghai Urban Space Art Season (SUSAS).

SUSAS inherits and carries forward the Expo 2010 spirit of "Better City, Better Life". Although it takes the form of an exhibition, it gives more connotation and value to urban space through the cooperation and interaction between the exhibition and urban renewal practices. The Space Art Season is held once every two years and consists of the main exhibition of space venue renovation, site projects, regional joint exhibitions, and public activities (SUSAS College). Each SUSAS is held to create a hot cultural topic, preserve the artworks made by masters at home and abroad, beautify the urban public space and make a "World Expo that never closes". Up to now, three exhibitions have been successfully held along the Huangpu River. Along with the construction of the "People's River" and the renewal process of the riverside area, these exhibitions have played an active role in the optimization and quality improvement of the spatial structure of the waterfront.

Both sides of the Pujiang River are closely related to Shanghai's urban development. With

浦江两岸与上海城市发展历程相戚相伴，拥有丰富的历史风貌遗存，是上海经济、社会、人文内涵丰富的展示窗口。20世纪末以来，随着城市产业结构调整、传统制造业转移以及内港外迁，浦江两岸的工业、码头功能已不适应时代发展需求，也出现了一系列消极空间的问题。因此，滨江地区的城市更新，一向是上海城市更新工作的重点和难点。因此，上海一方面积极推进滨江地区更新规划的编制实施，另一方面通过两年一度的"空间艺术季"，系统地推进和展示滨江城市更新和公共艺术实践，不断对滨江公共空间资源整治赋能。

从2015年室内展示规划成果和公共艺术品，到2017年实现工业遗产文化重塑，再到2019年走出场馆，以城市公共空间为载体，通过与城市更新和公共活动的深度融合，滨江"工业锈带"逐渐转变为"生活秀带"，浦江两岸作为上海全球城市功能品质标杆的形象逐渐清晰，空间艺术季活动也在内涵、深度、广度上不断拓展。三届主展分别选址于徐汇西岸飞机库、浦东民生码头8万吨筒仓、杨浦滨江毛麻仓库及滨江的5.5公里范围内。

徐汇西岸飞机库如今已成为民营美术馆，与西岸其他工业遗存改造的艺术场馆集聚形成了一条艺术品产业链走廊；浦东8万吨筒仓曾是亚洲最大的粮仓，极具历史和文化价值，结合贯通工程进行建筑改造，注入文化休闲、创意展示等多元功能，重新定义了滨江公共建筑和公共空间，在艺术季结束后已举办多场时尚、先锋艺术活动，成为深受欢迎的艺术空间；2019年的城市空间艺术季用上海船厂的两座巨大的船坞和具有百年历史的毛麻仓库作展场，再次带领观众领略空间和时间的魅力。杨浦滨江5.5公里的岸线上，每件公共艺术品都试图与场地融为一体，成为地景艺术，从静态装置展示到有动态参与的公共空间，实现了历史空间与现代人文的"相遇""对话"。可以说，空间艺术季活动通过对公共空间的赋能，将建筑空间改造、地区更新、视觉艺术设计和社会活动公众参与完整融合在一起，其本身就是现代城市更新理念的一次生动演绎。

以"艺术季"为触媒激发城市更新多重实践

"空间艺术季"作为城市经营的"文化事件"，不仅作用于空间艺术季展区所在地，其"触媒"和"标杆"作用更是激发了滨江45公里岸线的整体贯通和品质提升，其"生态、经济、人文"一体的空间赋能理念，也进一步带动了上海城市更新工作的全面展开和品质对标。

"滨江45公里贯通"打造世界级滨水功能带

根据黄浦江沿岸地区建设规划，两岸共十个主题区段，打造工业文明、海派经典、创意博览、文化体验、生态休闲、艺术生活等不同主题特色。杨浦滨江段利用老工业遗存更新改造，以工业传承为核心，打造历史感、生态性、生活化、智慧型的滨江公共空间岸线，滨江腹地新建大型办公、商业设施，完善地区功能，为周边市民服务。北外滩地区，位于虹口区南部，与陆家嘴、外滩隔江相望。作为"上海2035"国际航运中心功能

plenty of historical heritages, these places are the showcases for Shanghai's economy, society, and culture. Since the end of the last century, with the adjustment of urban industrial structure, the transfer of traditional manufacturing industries, and the relocation of the inner port, the functions of industries and docks on both sides of the Pujiang River have become unsuitable for the development needs of the times, and a series of space problems have emerged. Therefore, the renewal of the riverfront area has always been the focus and difficulty of Shanghai's urban renewal. Thus, Shanghai has been actively promoting the implementation of the riverside area renewal plan while displaying the achievements of renewal and public art practices through SUSAS, continuously contributing to the renovation of riverside public space resources.

From the indoor display of planning results and public artworks in 2015, to the realization of industrial heritage cultural reshaping in 2017, and then to the deep integration of urban public space with urban renewal and public activities in 2019, the riverfront "rust belt" has gradually transformed into a "show belt", both sides of the Pujiang River have gradually become the benchmark of Shanghai's global urban functional quality, and the activities of SUSAS have been more significant, sophisticated and inclusive. The main exhibitions of the three sessions were respectively located in the previous Longhua Airport airplane hangar of Shanghai West Bund Museum, Xuhui District, the 80,000-ton silo at Pudong Minsheng Port, the Maoma (linen and wool) warehouse with the 5.5-kilometer Yangpu waterfront public space.

The previous Longhua Airport airplane hangar of Shanghai West Bund Museum has now become a private art museum, forming an artwork industry chain corridor with other art venues converted from industrial relics of West Bund. The 80,000-ton silo at Pudong Minsheng Port was once the largest grain silo in Asia. It is of great historical and cultural value. After being restructured, the silo now can be a place for citizens to spend their leisure time and for creative display, making the riverfront public architecture and space more than its former usage. It has held several fashionable and pioneering art events after the art season, becoming a popular art space. The 2019 SUSAS uses the original Shanghai Shipyard's two huge docks and its 100-year-old Maoma warehouse as exhibition venues, once again leading the audience into the charm of space and time. In the 5.5-kilometer Yangpu waterfront public space, each public artwork tries to integrate with the site and become a landscape. From static display to a public space with dynamic participation, SUSAS realizes the "encounter" and "dialogue" between historical space and modern humanities. Through energizing the riverfront public space, SUSAS fully integrates architectural space renovation, regional renewal, visual art design, and public participation in social activities, being a vivid interpretation of the modern urban renewal concept.

"Art Season" as a catalyst to stimulate multiple urban renewal practices

As a "cultural event" of urban governance, SUSAS not only improves the area where the exhibitions locate but also serves as a "catalyst" and "benchmark" for the overall coherence and quality improvement of the 45km waterfront area. SUSAS promotes balanced ecological, economic, and humanistic spatial energizing progress and has further driven the comprehensive development of Shanghai's urban renewal and quality benchmarking.

"45 kilometers Joined Up Along the Riverside" to Create a World-class Waterfront Area

According to the construction plan for the Huangpu Riverfront area, there are a total of ten themed sections on both sides of the strait, creating different thematic features including industrial civilization, Shanghai classics, creative expo, cultural experience, ecological leisure, artistic life, etc.. The Yangpu Waterfront section makes use of the old industrial remains to renew and transform, with industrial inheritance as the core, to create a historical, ecological, life-oriented, and smart riverside public space. New large-scale office and commercial facilities are built in the riverside hinterland to improve the regional functions and serve the surrounding area. The North Bund area is located in the south of Hongkou District, facing Lujiazui and the Bund across the river. As the core bearing area of the "Shanghai 2035" international shipping centre, on the premise that the total building volume remains basically the same, it is committed to building the North Bund into one of the benchmarks of outstanding central activity areas of global cities. The Xuhui Riverside section, through a series of creative industry projects and the operation of cultural venues,

的核心承载区，在建筑总量基本维持不变的前提下，着眼更大区域联动，致力打造成为卓越全球城市中央活动区的标杆之一。徐汇滨江段，从煤码头和散货码头的集聚区，通过一系列创意产业项目和特色文化场馆运营，已成为全市文化创意产业集聚地；浦东滨江段规划建设世博文化公园，利用2010上海世博会场馆区域，总用地面积约188公顷。地块位于黄浦江转折的临江界面，堪称"小陆家嘴"，但为延续世博精神、提升中心城区空间生态品质，放弃了近千万平方米开发量，保留下宝贵的绿地，同时将文化与生态结合，配套建设上海大歌剧院等高等级文体设施。这些区域虽然分属不同行政区，但共同遵循"空间艺术季"传播的空间赋能理念，秉承"人民之江"建设总体目标，最终连点成线，共同打造全球城市"会客厅"和世界级滨水文化功能带。

"四大行动计划"引领城市公共空间更新活动

按照"卓越全球城市"的总体发展要求，除积极推进浦江贯通等重大公共空间建设外，上海也考虑市民的多样化活动需求，关注零星地块、闲置地块和小微空间的品质提升和功能创造，将"空间艺术季"理念引入社区，开展持续的城市更新"四大行动计划"，不断增加公共空间的面积和开放度，提高公共空间覆盖率和品质。其中社区共享计划关注老百姓的生活品质问题，重点改造社区消极空间；创新园区计划关注产业地区转型趋势，重点促进产城融合；魅力风貌计划关注历史文脉的保护传承难点，重点探讨历史遗存的活化利用；休闲网络计划则关注市民的健康和休闲需求，提供各种高品质公共空间场所。近年来城市更新项目已遍布全市，类型多样，包括复兴历史建筑和街区的长宁上生新所、静安新业坊；延续历史文脉的长宁愚园路、徐汇岳阳路、南汇新场古镇；改善社区消极空间、改造大量老旧小区的静安彭越浦河；产城融合的张江科学城、市北工业园区等。可以说，以空间艺术季为发端、以黄浦江滨江带为轴线，以"四大行动计划"铺展，上海已形成了不间断、全覆盖、高品质公共空间城市更新的良好局面。

以"共建共治共享"激励公共空间开发多重动力

在空间艺术季的策划和实施过程中，我们积极统筹政府、市场、市民三大主体，建立了贯穿全程的市场与公众参与机制，把空间艺术季的举办，演化成了引导全社会共建、共治、共享城市更新成果的实践过程。

政府层面

市级政府统筹把控整体建设方向。由市领导牵头，市级各相关部门和各区政府组成联席会议机制，出台《关于提升黄浦江、苏州河沿岸地区规划建设工作的指导意见》作为规划建设纲领性文件，共同研究制定相关政策，讨论审议重要地区规划建设方案，协调解决重大问题。由市级规划资源部门深化城市设计专项研究，加强对天际轮廓线、色彩、公共空间、地下空间等方面的全局管控；将相关要求落实到附加图则中；细化建管要求，充分运用三维审批等手段，加强重点区域项目管理，为沿

has transformed from a gathering area of coal wharf and bulk cargo wharf to the city's cultural and creative industry gathering place. It is planned to build the World Expo Cultural Park on the basis of the 2010 Shanghai World Expo venues along the Pudong Riverside section, with a total land area of about 188 hectares. The turning sector of the Huangpu River, being called "Little Lujiazui", abandons further development and preserves the green land in order to carry on the Expo spirit and improve the ecological quality of the central urban area. Under the tenet of culture integrating with ecology, high-grade cultural and sports facilities such as the Shanghai Grand Opera House have been built in this sector. Although these areas belong to different administrative districts, they jointly follow the SUSAS concept of energizing the city with space, uphold the overall goal of constructing a "People's River", and finally together create a global city "parlor" and world-class waterfront cultural belt.

"Four Action Plans" to Lead Urban Public Space Renewal

Under the overall development requirements of the "Global City of Excellence", in addition to actively promoting the construction of major public spaces such as the Pujiang Riverside, the diversified activity needs of citizens should also be considered. Shanghai pays attention to the quality improvement and functional creation of fragmented plots, unused plots, and small and micro spaces, introduce the concept of "Space Art Season" into communities, and carries out the "Four Action Plans" for continuous urban renewal, to continuously increase the area and openness of public space and improve its coverage and quality. Among the four plans, the Community Sharing Plan focuses on people's life quality and renovates negative community spaces; the Innovation Park Plan is centred on the transformation trend of industrial areas and promotes the integration of industries and cities; the Charming Landscape Plan concentrates on the difficulties of preserving and passing on historical remains and explores the revitalization and utilization of historical relics; the Leisure Network Plan is targeted on the health and leisure needs of the public and provides various high-quality public places. In recent years, urban renewal projects have been spread all over the city with various types, including the revitalization of historical buildings and neighborhoods in Changning Columbia Circle and Jing'an Xin Ye Fang; the continuation of historical heritage in Changning Yuyuan Road, Xuhui Yueyang Road, and Nanhui Xinchang Ancient Town; the improvement of negative community space and transformation of a large number of old neighborhoods in Jing'an Pengyuepu River; the integration of industry and city in Zhangjiang Science City and Industrial Park in the North Shanghai, etc. With the Space Art Season as a starting point, the Huangpu River waterfront as the axis, and the "Four Action Plans" as a strategy, Shanghai has formed a good situation of uninterrupted, full-coverage, high-quality public space urban renewal.

"Co-construction, Co-governance and Sharing" to Stimulate Multiple Dynamics of Public Space Development

In the planning and implementation of the Space Art Season, we have actively coordinated the three main bodies of the government, the market, and the citizens, established a market and public co-participation mechanism throughout the whole process, and developed the art season into a practical process of guiding the whole society to the co-construction, co-governance and sharing of urban renewal.

The Government

The municipal government coordinates and controls the overall construction direction. Led by municipal leaders, all relevant municipal-level departments and governments at the district level form a joint meeting mechanism, jointly study and formulate relevant policies, discuss and consider plans for important areas, and coordinate and resolve major issues. The *Guidance on Enhancing Planning and Construction Work in Areas Along Huangpu River and Suzhou Creek* has been issued as a programmatic document. The municipal-level planning and resource departments pay efforts to deepen special studies on urban design, strengthen control on skyline contours, colours, public spaces, and underground spaces, and implement relevant requirements into additional plans; refine construction and management requirements, make full use of technical measures such as 3D approval to strengthen project management in key areas and provide guarantees for the quality of spaces along the shoreline. The district-level government takes the lead

岸建筑空间品质提供保障。由区级政府主导推动实施协调，明确实施项目、实施主体、实施策略和时间要求，保障规划有序实施。

市场层面

充分调动企业积极性，引导多元主体共同参与建设。为引入多元主体共同参与开发，上海推出一系列适应市场的城市更新政策。比如打破区政府收储进行改造的单一路径，以释放更多的公共设施和公共空间为前提，鼓励物业权利人按规划进行更新改造；比如实施"带方案"招标挂牌复合出让，引导和激发市场主体不断提高设计和建设品质。在实施过程中始终算好空间账、经济账和时间账，通过科学的开发时序和目标统筹，平衡好滨江第一层面开发和腹地开发之间的互动，协调好城市历史遗存保护和活化利用的价值，处理好当前投入和长远收益的关系，将资金和市场共同引导到政府对滨江地区长远发展的战略理念上来。

社会层面

引入多样化的公众参与空间治理机制。改变传统规划资源管理理念，在公众参与的深度上，不再仅限于规划编制阶段草案的公示，而是贯穿城市规划管理实施的全过程，公众不再局限于被动的意见征询，而是通过定期举办城市空间艺术季这样的特色活动，让市民感知并参与到城市的更新改造中；在公众参与广度上，公众参与的形式也是多种多样的，比如激发社区自治开展的微更新，发动全民为上海的城市更新建言献策；比如建立社区规划师制度，引入专业力量扎根社区；比如与社区管理联合，在公共空间中设立党建服务站，鼓励市民参与公共空间志愿者服务等，让广大市民在城市空间的建设、使用、管理过程中，获得与城市共建共治共享的参与感和成就感。

结语

近六年来，空间艺术季活动已取得了一定的影响力，办展主旨及独特的形式，得到了国内外业界的高度关注和十分积极的评价，已初步形成了一定的城市品牌效应。在今后的城市更新的实践过程中，上海将继续举办好空间艺术季，一是进一步打造城市空间新样板，推动"人民城市"建设。以改善和提升空间品质为根本深入开展城市更新，为人民群众不断创造美好的城市公共空间和生活环境，提高公众参与度及覆盖面，增强群众获得感；二是进一步提供城市发展新动能，放大活动乘数效应。统筹好政府、社会、市民三大主体，调动各方参与城市有机更新的积极性，以城市空间品质提升和文化艺术事件经营的叠加形成乘数效应，促进区域经济和文化密度的提高；三是进一步拓展展览活动新主题，扩大活动辐射影响力。创新工作思路，丰富活动形式和内容，在活动主题上，将从滨水空间逐步拓展到其他各类公共空间；在活动区域上，也将从城市中心地区逐步拓展更贴近市民的社区以及郊野乡村地区。不断提升空间艺术季的内涵，扩大活动的深度、广度和影响力，为建设更有魅力、更有活力、更有温度的人民之城，贡献规划资源的智慧和力量。

in promoting implementation and coordination, clarifying the projects, implementation subjects, implementation strategies, and time requirements, and ensuring the plan is carried out orderly.

The Market

Fully activate enterprises and lead all levels of the society to participate in the construction together. Shanghai has introduced a series of urban renewal policies that are adapted to the market. For example, stop relying only on government acquisition and storage for reconstruction, release more public facilities and public space and encourage property right holders to carry out renewal and renovation according to the plan; and the scheme-based bid-quotation assignment approach which aims to guide and stimulate market agents to continuously improve their design and construction quality. In the implementation process, fully take the spatial, economic, and time cost into account. Through scientific development sequence and target coordination, balance the interactions between riverfront open spaces and hinterland developments, coordinate the protection and utilization of urban historical heritage, deal with the relationship between the current investment and long-term benefits, make the capital and the market serve for the government's strategy of the riverfront area long-term development.

The Society

Introduce diversified mechanisms for public participation in space governance. Change the traditional concept of planning resource management. Public participation should no longer be limited to the publicity of the draft plan at the planning stage but runs through the entire process of implementation. Apart from being consulted a passively, citizens can perceive and participate in the renewal and transformation of the city through special events such as SUSAS. Public participation should be achieved in various ways, such as encouraging the community to make micro-renewal, mobilizing citizens to offer suggestions for Shanghai's urban renewal; establishing a community planner system, and introducing professionals to work in the community; in cooperation with community management, setting up service stations of the Party in public spaces, encouraging citizens to participate in public space volunteer services, etc. As a result, the public can possess a sense of participation and accomplishment of co-construction, co-governance, and sharing in the process of space construction, use, and management.

Conclusion

In the past six years, SUSAS has become quite influential. The main purpose and special format of the exhibition have received high attention and praises from the art industry at home and abroad, and have initially created a certain urban brand effect. In the future, we will continue to make efforts to hold each SUSAS successfully. Firstly, to create a new model of urban space, and promote the construction of "people's city". We will improve and enhance the quality of space as the fundamental to carry out in-depth urban renewal, continue to create a better urban public space and living environment for the people, increase public participation and its coverage, and enhance the people's sense of gain. Secondly, to provide new momentum of urban development and amplify the multiplier effect of activities. Coordinate the government, society, and citizens, mobilize all parties of the society to participate in the city renewal. Form a multiplier effect with the quality improvement of urban space and cultural and artistic events, to promote the improvement of regional economic and cultural density. Thirdly, to develop new themes of exhibition activities and increase the influence. Be more creative at work and enrich the form and content of activities. The focus of our exhibition will gradually transform from the waterfront space to other types of public space, while the area of activities will also be expanded from communities close to the public to the countryside rural areas. We will continue to improve the connotation of the space art season, hold various activities with more depth, breadth, and influence, and make full use of our planning resources to build a more attractive, more vibrant, and warmer city for the people.

序 2
打响上海文化品牌 助推城市品质提升

Preface 2
Build Shanghai's Cultural Brand, Boost Urban Quality Improvement

上海市文化和旅游局
Shanghai Municipal Administration of Culture and Tourism

文化是城市的灵魂，上海是文化的"码头"，更要做文化的"源头"。设立城市空间艺术季，既是在源头上不断激活和推动文化创新，又是展现城市品位、提升城市品质的重要举措。创立5年来，空间艺术季将创造和谐的人居环境、营造宜人的艺术氛围、以城市更新助力文化传承与创新、推动城市公共艺术发展作为目标，已逐渐形成了以激活城市空间的艺术综合展演为形式表征，以关注日常的人文精神表达为美学特征，以深度参与的交互互动为亲民体验特点，日趋成为新的上海文化品牌。

本届空间艺术季以杨浦滨江为主展场，既是落实《全力打响"上海文化"品牌加快建成国际文化大都市三年行动计划（2018—2020年）》中关于"实施'海派城市地标'品质提升计划，建成特色鲜明的滨江文化长廊"的工作要求，也是基于城市自身发展进程的客观需要。见证上海由近现代民族工业发展的城市工业遗存转换为新的文化综合体，需要从内容的建构上注入更为丰盈的时代文化艺术元素。在前两届的基础上，本届空间艺术季在整体内容架构、公共艺术作品设计、展陈形式、关注观众体验等各个方面都呈现出新的亮点。

艺术之美，塑造城市个性魅力

空间艺术季的内容架构是上海文化发展视野的脚注，既有基于杨浦区域特质、注重文脉梳理的作品，又不乏国际视野的活力展现。开幕展演的合唱套曲《相遇》以个体生命为着眼点，关注历史文脉的梳理，充满人文关怀的温度。作品由七首单曲组成，按时序描绘了杨浦滨江的七个片断，既是一天之中的七个缩影，也代表着杨浦滨江的前世今生；由烟草仓库化生的"绿之丘"多元空间，展示"杨浦七梦"，围绕杨浦的七个关键词"体育、工人、教育、音乐、河流、纺织及消遣"，勾勒出曾经生活、工作在此地的人们，发生过怎样的故事，如今又对此怀有怎样的梦想。而上海街头艺术节开幕系列杨浦专场和世界舞蹈大赛（中国）巡演，则充分体现上海文化发展的国际视野，前

Culture is the soul of the city. Shanghai is the "dock" of culture, and more importantly, it should be the "source" of culture. Holding the Urban Space Art Season is not only to fundamentally activate and promote cultural innovation, but also an important measure to display and improve the cultural environment of the city. In the past five years since its establishment, the Space Art Season has targeted to create a harmonious living environment, forge a pleasant artistic atmosphere, advocate cultural inheritance and innovation with urban renewal, and support the development of urban public art. It has gradually become a new Shanghai cultural brand with a form of a comprehensive art exhibition that activates urban space. It has been dedicated to delivering a kind of daily spiritual culture and enabled the public to enjoy deep participation and interaction with art.

This year's SUSAS takes the Yangpu Riverfront as the main exhibition venue. The *Three-year Action Plan (2018-2020) to Speed Up the Building of the Brand of "Shanghai Culture" and an International Cultural Metropolis* makes the requirement to improve the city quality of Shanghai as a landmark. SUSAS 2019 is not only an endeavor to meet the up-mentioned requirement, but also based on the objective needs of the city's development. In order to show the transformation of Shanghai from the urban industrial remains left by modern national industry development to a new cultural complex, this art season needs more cultural and artistic elements of the times from the construction of content. Based on the previous two sessions, this year's SUSAS presents new highlights in the overall content structure, public artwork design, exhibition format, and audience experience.

The Beauty of Art, Shaping the City's Personality

SUSAS is a showcase of Shanghai's cultural development vision, with works both based on Yangpu's regional characteristics and combining with the history and the international perception. The song cycle *Encounter* is displayed at the opening ceremony. It focuses on individual lives and the combing of history, and culture, and is infused with humanity. The work consists of seven songs, depicting seven scenes of Yangpu Riverfront in chronological order, which are the snapshots of the day, representing the past and the present of Yangpu Riverfront. The Green Hill, a space developed from the original tobacco warehouse, showcases the "Yangpu Seven Dreams". It is designed around seven keywords of Yangpu: sports, workers, education, music, river, textile, and leisure, depicting the stories of people who used to live and work here and the dreams they have today. The opening series of the Shanghai Street Performance Festival and the World of Dance (WOD) China Tour fully reflect the international vision of Shanghai's cultural development. The Street Art Festival brings together street art masters from more than 10 countries and Shanghai artists to exchange their skills in Shanghai Fashion Centre, while WOD brings "the dancing world" into the public life, connecting people of all ages and backgrounds through dancing. These dynamic art performances together with the static public sculpture exhibitions mobilize the audience's multiple senses to feel and experience art in an all-around way, invisibly shaping

者汇集十多个国家的街艺高手与上海艺人在上海国际时尚中心一起交流技艺，营造台上台下欢乐互动的节庆气氛；后者将"舞蹈世界"融入大众的生活，通过舞蹈语言将不同年龄和背景的人们联系在一起。这些动态的艺术展演与静态的公共雕塑作品展相融合，构成动静相宜的节奏，调动观众的多重感官全方位感受与体验艺术，无形之中塑造着城市的个性与魅力。

以人为本，共享城市艺术活力

以人为本，体现在作品以人的情感、人的经历为表现主体，还体现在将观众的参与和互动作为重点。作为文旅融合发展后举办的第一届空间艺术季，通过艺术项目丰富和提升上海旅游的内涵和品位，通过旅游人气扩大空间艺术季的辐射面，相辅相成。主办方设计了"工业遗存体验之旅"线路，观众从本届空间艺术季的主展馆上海国际时尚中心出发，徒步滨江公共空间，探寻工业遗迹变身时尚地标，听建筑和雕塑背后的故事。此次活动还招募百名市民参与艺术家地绘工坊，在为期1个月的时间里，创作出长达200米的艺术作品《城市的野生》，永久保存在杨浦滨江的开放空间中。同时，在杨浦滨江5.5公里区段内，共有22处文物和历史建筑可以扫码听故事，还开通了"智慧导览感应系统"，12个感应点位陪伴市民观光讲解导览，带来多维度沉浸式的游览体验。

上海城市空间艺术季活动将进一步围绕"人民城市"理念，紧扣时代发展的脉搏、紧跟城市发展的脚步，通过持续不懈的创意和创新，不断丰富和提升品牌内涵，为人民创造更多、更美的艺术空间，组织更丰富、参与性更高的文化活动，满足人民群众对美好生活的向往，打造独具魅力和特色的文化品牌。

the personality and charm of the city.

People-oriented, Sharing the City's Artistic Vitality

The people-oriented approach is reflected in the fact that the works take the emotions and experiences of mankind as the main expressions and the participation and interaction of the audience as its priority. This year's space art season is the first one held after the integration of culture and tourism development. The connotation and taste of Shanghai's tourism are enriched by the art projects while the influence of space art season is expanded through tourism popularity. The audience can take the "Industrial Heritage Experience Tour", walk along the riverfront public space starting from the Shanghai Fashion Centre, which is the main exhibition hall of this year's SUSAS, explore the fashion landmarks transformed from industrial relics and listen to the stories behind the buildings and sculptures. The event also recruits 100 citizens to participate in an artist's ground painting workshop to create a 200-meter-long artwork *Wildness Growing Up in the City* over one month. This artwork will be permanently preserved in the open space of the Yangpu Riverfront. In the 5.5 km Yangpu Riverfront section, the audience can scan the code to listen to the story of 22 cultural relics and historical buildings. A smart guide system, with 12 induction points for the public sightseeing tour guide, has also been introduced to create a multi-dimensional immersive tour experience.

The Shanghai Urban Space Art Season will further focus on the concept of "People's City", closely follow the times, and keep up with urban development. Through unremitting creativity and innovation, we will continuously enrich and enhance the brand connotation, create more and better art space for the people, organize various cultural activities to satisfy the aspirations of the people to live a better life and create a cultural brand with unique charm and characteristics.

滨水空间与公共艺术的结合——城市"展场"的探索

The Combination of Waterfront Space and Public Art—An Exploration of Urban "Exhibition Venue"

郑时龄
2019 上海城市空间艺术季
学术委员会主任

Zheng Shiling
Director of SUSAS 2019 Academic Committee

杨浦滨江环境的整治早在 2010 年就已经开始讨论，2016 年正式启动，2017 年实现了滨江带的贯通。2019 上海城市空间艺术季以杨浦滨江南段作为主展场，相当于将之前的贯通工程向杨浦大桥以北的下游进行了延伸。与其他区域相比，杨浦滨江有更多工业遗存，包括厂房、仓库、码头等，是一种相对硬质化的空间。但这里的城市化程度较低，空间具有多元化的潜力。

上海城市空间艺术季旨在挖掘和利用上海的各处公共空间。虽然大家的目光主要集中在室内的展场，但城市空间的更新与利用逐渐被置于重要的位置。"西岸 2013 建筑与当代艺术双年展"在徐汇滨江举办后，2015 年举办了第一届上海城市空间艺术季。之后每隔两年举办一届，从徐汇西岸飞机库旧址到浦东民生码头 8 万吨筒仓旧址，再到杨浦滨江从秦皇岛路到定海路段、岸线长约 5.5 公里的滨江公共空间，每一届的展览地点都不同。不过这些地区具有一个共同的特征——都拥有正在变化和发展的滨水空间。2019 年的艺术季相当于对上海城市滨水空间的发展进行了总结，下一届或许会转向其他方面，比如城市中其他类型的区域，或者郊区。

将公共艺术植入城市空间

随着上海的经济与社会发展，将公共艺术作品植入城市空间，是对不断增大的文化需求的回应。2015 年的艺术季的展陈项目主要是在规划与建筑领域。2017 年开始加入更多艺术作品，原本制定了公共艺术的植入计划，但由于条件的限制，没有来得及实施。2019 年终于将公共艺术布置在杨浦滨江 5.5 公里的展场上，艺术作品深入到城市中，与公共空间紧密地融合，空间因此变得更加活跃。

The improvement of the Yangpu Waterfront environment has been discussed as early as 2010, and it was officially launched in 2016. In 2017, the construction of the riverfront belt was completed. The 2019 Shanghai Urban Space Art Season takes the southern section of Yangpu Riverfront as the main exhibition venue, which is equivalent to extending the riverfront belt to the downstream north of Yangpu Bridge. Compared with other sections, Yangpu Riverfront has more industrial relics, including factories, warehouses, and wharves, which is a relatively rigid space. However, it is not highly urbanized here and has the potential for diversification.

SUSAS aims to explore and utilize various public spaces in Shanghai. For a long while, indoor exhibition venues have been the focus of most people. The renewal and utilization of urban space are now gradually being attached with great importance. After the "West Bund 2013: A Biennial of Architecture and Contemporary Art" was held in the Xuhui Riverfront area, Shanghai held the first Urban Space Art Season in 2015. Since then, SUSAS has been held every two years. Each session has chosen different locations, from the original site of the aircraft hangar of Xuhui West Bund to the old 80,000-ton silo at Pudong Minsheng Port, and to the public space along the Yangpu Riverfront from Qinhuangdao Road to Dinghai Road, with a shoreline of about 5.5 km. However, these areas share a common characteristic - they all have waterfront spaces that are changing and evolving. The 2019 art season is a summary of the development of Shanghai's urban waterfront spaces, and the next session may turn to other aspects, such as other types of areas in the city or the suburbs.

Implanting Public Art into Urban Spaces

With the economic and social development, the implantation of public artworks into the urban space is a response to the ever-growing cultural demand. The exhibition projects of the 2015 SUSAS are mainly in the field of planning and architecture. In 2017, more works of art started to be introduced. There was supposed to be a public art implantation plan but failing to carry out due to conditions. In 2019, we finally make it real: the 5.5-kilometer exhibition area of Yangpu Waterfront for public art. The artworks are deeply integrated with the city, and thus making the public space more active.

Public art is relatively common in the world. It is an art form facing the public and is significant in urban space. Public art attempts to communicate with the environment, rather than merely expressing the artist's ideas. High-density cities usually need iconic urban space. Public art can be used to build a landmark, create a place recognized by the public,

公共艺术在国际上比较普遍，是一种面向公众的艺术形式，在城市空间里拥有一席之地。公共艺术试图跟环境发生对话，而不是仅仅表达艺术家的想法。高密度城市通常需要具有标志性的城市空间，可以通过设置公共艺术来建立区域的标志性，创造一个受公众认可的场所，并给人们留下属于这里的记忆。一旦大家都汇集到这里，就能使空间生出新的活力。

上海城市空间艺术季跟通常的艺术展览之所以不同，是因为它试图借助展览来促进城市的发展和更新。学术委员会十分期望2019年参展的公共艺术作品能够跟城市的环境高度契合，成为一种只有在这个地方才能形成的"唯一"的艺术。这相当于对艺术家也提出了更高的要求。艺术家需要事先研究上海的文化环境，要与建筑师配合，还要充分利用场地的特点。原来船厂遗留的船坞、仓库、码头上的系船柱，甚至是杨浦大桥，都可以成为艺术作品的一部分。杨浦大桥本身也带有一种雕塑性，仿佛城市空间中的一件巨大的装置艺术。公众会因此而相信，这些艺术作品是属于大家的。

学术委员会与策展人的作用

2019年的艺术季学术委员会委员十分多元，既有建筑师、城市规划师，也有艺术家、策展人，甚至还有法律方面的专家，再加上上海城市公共空间设计促进中心的协调，在各方之间起到润滑和联系的作用，学术委员会得以发挥出真正的作用。委员们主要讨论展览的内容，包括对艺术作品的评价和选择，大家从各自的专业出发，提出不同的想法。策展人在艺术家的选择上起到决定性作用，所以委员会对策展人的选择也经过了多轮讨论。

在策划前期，策展人根据艺术家的自身特点以及环境所提出的要求，来选择适合的参展艺术家。可能具体到展场中的每个点位，各个点位都有多组相应的人选。策展人把城市空间艺术季的目标和宗旨传达给艺术家，根据他们计划提出的作品，来判断是否跟环境契合，并最终做出选择。

"相遇"包含了一种期待

2019城市空间艺术季的主题"相遇"，是让人们与城市的历史、城市的今天、城市的未来相遇。"相遇"包含了一种期待，期待人们走近滨水空间，感受滨水空间的魅力，所以"相遇"也蕴含了人与人的相遇、人与空间的相遇、人与水的相遇、艺术与城市的相遇……上海黄浦江两岸的滨水空间包含历史的、文化的、社会的意义，通过多种多样的"相遇"，这些意义或许可以获得一次新的升华。

and leave people with memories of this place. Once people are gathered here, new vitality can be created in the space.

SUSAS is different from the usual art exhibitions because it tries to use exhibitions to promote the development and renewal of the city. The Academic Committee very much hopes that the public artworks exhibited in 2019 will be highly compatible with the urban environment and become the art that can only be formed in this place. This brings higher requirements to the artists. The artists need to study the cultural environment of Shanghai in advance, cooperate with the architect and make full use of the characteristics of the site. The docks, warehouse, bollards on the docks and even the Yangpu Bridge left by the original shipyard can all become part of the artwork. The Yangpu Bridge itself can be viewed as a sculpture, like a huge installation artwork in an urban space. The public will therefore believe that these works of art belong to everyone.

The Role of the Academic Committee and The Curator

The members of the Academic Committee for the 2019 art season consist of people from all aspects. We have architects, urban planners, artists, curators, and even law experts. With the coordination of the Shanghai Design & Promotion Centre for Urban Public Space, which acts as a lubricant and link between all parties, the Academic Committee comes into real play. The committee mainly focuses on the content of the exhibition, including the evaluation and selection of artworks. Members bring up different ideas from their respective professions. The curators play a decisive role in the selection of artists, so the committee has thought carefully about the selection of curators before making the decision.

In the early stage of planning, the curators select suitable participating artists based on the artist's characteristics and the environmental requirements. The selection may be specific to each point in the exhibition hall, and each point has multiple sets of corresponding candidates. The curators convey the goals and objectives of our art season to the artists and estimate whether they are compatible with the environment according to the works proposed, and finally make a choice.

"Encounter" is an Expectation

The theme of the 2019 urban space art season, "encounter", is to let people meet the history, the present, and the future of the city. "Encounter" is also an expectation, expecting people to get close to the waterfront space and feel the charm. It implies the encounter between different people, people and space, people and water, art and the city, etc.. The waterfront spaces on both sides of the Huangpu River in Shanghai are historically, culturally, and socially important, and can be revitalized via "encounters".

一种发展的、包容的展览策略
A Developmental and Inclusive Exhibition Strategy

丁乙
2019 上海城市空间艺术季
学术委员会委员

Ding Yi
Member of SUSAS 2019 Academic Committee

2015 年第一届艺术季中，我受邀在户外进行了些装置作品的创作。我当时深受这个展览形式的触动，因为它打破了通常意义中发生在美术馆的艺术展，而是选择聚焦建筑、艺术、空间互相交错的多元跨界呈现。

上海城市空间艺术季每届都选择一个不同的展览场地，可以说是一种游离式或者移动式的展览。2015 年在徐汇滨江，2017 年在浦东，2019 年在杨浦滨江，从这三个地点可以看到，艺术季好像是一个不断延伸和发展的展览。

上海城市空间艺术季还遵循边改造、边呈现的模式。当下城市中大大小小的展览，多是按照美术馆或展览中心的体量或架构来策划的。而艺术季是在场地还没有完善的前提下，就介入其中。比如第一届的主场馆西岸艺术中心当时还没有发展成熟，如今已经成为上海的重要美术馆之一。第二届的主场馆民生码头曾被废弃许久，通过举办展览进行了改造。2019 年不仅将毛麻仓库改造为主展馆，而且 5.5 公里的杨浦滨江区域也成为主要室外展场。

应创造怎样的公共艺术？

公共艺术是一个很小、也很大的概念。从西方传入中国，近几年，越来越多中国人开始从事公共艺术相关的研究和教学。我在上海视觉艺术学院任教的时候，曾创立公共艺术专业，当时还没被列在教育部的学科目录里，归于美术学院雕塑专业。如今已被列入学科目录，未来可能有越来越多的专业人才在全国各地培养出来。

公共艺术首先是互动的。传统的雕塑仅仅是艺术家的自由创作，全按自己的风格来实现，然后释放到城市的一个点位上或者一个空间里。公共艺术与之不同，它更关注与其所处空间的结合。因此，一个优秀的公共艺术作品往往都是有针对性的。在前期，艺术家要要先了解作品所在区域的文化、历史、甚至是该区域和周边的关联，然后从中提取一些因素，进行适合这个区域的创作。

In the first space art season in 2015, I was invited to create some installation works outdoors. I was deeply touched by the format of this exhibition at that time, as it broke away from the usual art exhibitions that took place in art galleries, and instead chose to focus on multiple cross-border presentations where architecture, art, and space intersect with each other.

The Shanghai Urban Space Art Season selects a different venue for each session, which can be described as a kind of free or mobile exhibition. The venue of SUSAS 2015 was in the Xuhui Riverfront area, 2017 in Pudong, and 2019 in the Yangpu Riverfront area. Through this, it seems that the art season is constantly extended and developed.

SUSAS follows the model of transforming and presenting as it goes. Most of the exhibitions in the city today, large and small, are planned according to the volume or structure of an art museum or exhibition centre. SUSAS, on the other hand, starts its planning before the venue is perfected. For example, the main venue of the first session, the West Bund Museum, was not yet developed and has now become one of the major art museums in Shanghai. The second session's main venue, Pudong Minsheng Port, had been abandoned for a long time and was renovated by holding exhibitions. SUSAS 2019 not only transforms the Maoma warehouse into the main exhibition hall but also makes the 5.5 km Yangpu Riverfront area the major outdoor venue.

What Kind of Public Art Should Be Created?

Public art can be a concept both small and big. It was introduced to China from the West, and in recent years, more and more people in China have begun to engage in public art research and teaching. When I was a teacher at the Shanghai Institute of Visual Arts, I founded the major of public art, which was not yet listed in the discipline catalog of the Ministry of Education. As a result, it was under the sculpture major of the Academy of Fine Arts. Now that it has been included in the discipline catalog, more and more professionals may be trained nationwide in the future.

Public art is, above all, interactive. The traditional sculpture is simply the free creation of an artist in his or her style and then laid somewhere in the city. Public art is different. It is more concerned with the integration of the space in which it is located. Therefore, good public artwork is often targeted. In the early stages, the artist must first understand the culture and history of the area where the work will be placed, and even the connection with the surroundings, and then extract some factors from it to create something suitable for the area.

In the 2019 session, in order to create works that evoke the memories of citizens, many artists have conducted background research on Yangpu Waterfront and Shanghai. The organizing committee also gives their full support to observe the environment, and collect relevant materials, such as the equipment in the original Yangpu Waterfront factory, etc..

在 2019 年的艺术季中，很多艺术家都对杨浦滨江以至上海市进行了背景调研，组委会也陪同他们去考察环境，甚至是收集和当地有关的材料，比如杨浦滨江原来厂区里的器械、物件等，从而创作出勾起市民回忆的作品。对杨浦滨江来说，这些公共艺术延续着过去工业时代社会发展的痕迹，人们对往昔的回忆与艺术相结合。2019 年的艺术季共有 20 件永久艺术作品被设置在杨浦滨江。实际上，虽然上海的艺术活动一直非常活跃，但如此高水准的、国际性的、真正属于城市的永久性公共艺术作品非常少。

上海的城市特质

上海这座城市的最大特征是国际化。任何一个在上海生活过的人，都会有一些类似的记忆，比如偶遇散落在市区中的老洋房。那些房子多建于 20 世纪 30 年代，正是上海人口剧增、城市建设突飞猛进的时期。那时的建筑风格主要是装饰艺术风格（Art Deco），最早在 1925 年巴黎装饰艺术博览会上出现，而上海 1927 年就有此风格，其国际化程度可见一斑。

上海不仅始终把眼光放在世界上最先进、最优异、最流行的层次上，还追求一种特殊的品质。这种品质既包含中国文化基因的传承，又包含放眼世界的眼光和胸怀。因此可以说，上海是一个海纳百川的城市。这跟上海的历史遗存和文化因子有关，比如我工作室所在的地方是中国民航发源地之一，岸边还留着许多遗留的民族工业。同样的历史在杨浦也有很多，包括具有上海特色的纺织厂、船厂，代表了上海的工业发展。这种本土文化与外来技术的碰撞和融合，造就了上海多元、丰富的气质。

"相遇"的包容性

"相遇"充满着包容性，可以是不同时空、不同地域的艺术家和上海的相遇，可以是市民和艺术的相遇，也可以是不同专业的人们在杨浦滨江上的相遇，如艺术家、建筑师、城市规划师和工人们。这个主题融合了许多不同的方面，具有宽阔的时空感，几乎可以说是一次关于历史和当代艺术的碰撞。相遇在这样的地点可能会产生十分震撼的感觉，或者被勾起回忆，唤起联想。

一开始听到"相遇"这个主题的时候，我认为它比较宽泛。在采访了很多人之后，我意识到这个主题反而具有上海特性。因为上海从发展伊始即保持着开放、包容的姿态，各种各样的文化交融于此，所以如果结合上海的特质来思考"相遇"，会发现它像一个容器，可以放入很多东西，包括个人的命运、城市的命运，甚至是国家的命运。其中也有很多点可以继续延伸下去，产生新的可能性。而且，公共艺术的特点之一也是要为更大范围的受众创造连接的契机，这一点也与"相遇"不谋而合。

Traces of social development in the industrial era can still be found in the public artworks along Yangpu Riverfront. Memories of the past are combined well with art. A total of 20 permanent artworks have been set up on the Yangpu Waterfront in the 2019 SUSAS. Honestly, although Shanghai's art activities have always been very active, there are very few permanent public artworks that can reach such a high level, international, and truly belong to the city.

The Characteristics of Shanghai

The biggest feature of Shanghai is its internationalization. Anyone who has lived in Shanghai must have memories like coming across the old western-style houses in the urban area. Most of those houses were built in the 1930s when the city was booming with increasing population and progressive urban construction. At that time, the architectural style was mainly Art Deco, which first appeared at the Art Deco Fair in Paris in 1925. And this style started to appear in Shanghai in 1927. It's not hard to tell how international Shanghai is.

Shanghai always strives to be one of the most advanced, the best, and the most popular in the world, and at the same time pursues a special quality. This quality contains both the inheritance of Chinese culture and a kind of global vision. So, Shanghai is a city that embraces all. The historical relics and cultural elements have something to do with this. For example, the place where my studio is located is one of the cradles of Chinese civil aviation, and there are many national industrial factories left on the riverbank. We can find a lot of similar places in Yangpu, such as textile mills and shipyards with Shanghai characteristics which represent the industrial development of the city. The collision and fusion of local culture and foreign technology have made Shanghai a diverse and distinctive metropolis.

The Inclusiveness of "Encounter"

"Encounter" is inclusive. It can be the encounter between Shanghai and artists from different times and areas, between citizens and art, or between people from different professions, such as artists, architects, urban planners, and workers. This theme blends many different aspects with a broad sense of time and space. It is a collision of history and contemporary art. The encounter in such a place can be very impressive and evocative.

When I first heard about the theme of "encounter", I thought it was relatively broad. After interviewing many people, I realized that this topic actually has Shanghai characteristics. Shanghai has been open and inclusive since the beginning of its development and different cultures have mingled here. If you think about "encounter" in combination with the characteristics of Shanghai, you will find that it is like a container in which you can put many things, including the destiny of the individual, the city, and even the country. There are also many other possibilities. What's more, public art is also an opportunity to create connections for a wider range of audiences, which is another kind of "encounter" as well.

策展团队
Curatorial Team

总策展人
Artistic Director

北川富朗
Fram Kitagawa

日本新泻县高田市人，国际知名艺术策划人。日本"越后妻有大地艺术祭""濑户内国际艺术节"发起人、艺术总监；东京"FARET立川"公共艺术总体策划人。

Fram Kitagawa was born in Niigata Prefecture, Japan. He is the founder and artistic director of "Echigo-Tsumari Art Triennial" and "Setouchi International Art Festival" and the artistic director of "Faret Tachikawa" public art.

规划建筑版块策展人
Architecture and Urban Curator

阮昕
Ruan Xin

上海交通大学设计学院院长、光启讲席教授、校学术委员会委员。曾任澳大利亚新南威尔士大学城市建筑环境学院副院长、建筑系主任（2004—2018），悉尼科技大学建筑学院院长（2002—2004）。现任国际建筑师协会建筑评论家委员会（CICA）委员、美国建筑历史学家学会（SAH）会员、澳大利亚建筑师学会（AIA）学术会员、新南威尔士大学建筑学长聘教授、同济大学顾问教授。

Ruan Xin, Ph.D, is Dean and Guangqi Chair Professor of Architecture at School of Design, Shanghai Jiao Tong University. He is a member of Shanghai Jiao Tong University Academic Committee. He joined UNSW Sydney as Professor of Architecture in 2004. He was Associate Dean International (2015-2018), Director of Architecture (2014-2016), Chair of Architecture Discipline and Director of Master of Architecture from (2005-2009). Prior to his appointment to UNSW, he was Director of School of Architecture at the University of Technology Sydney (2002-2004).

空间艺术版块策展人
Urban Space Art Curator

川添善行
Yoshiyuki Kawazoe

建筑家、东京大学副教授。任空间构想一级建筑士事务所代表、日荷建筑文化协会会长等。代表作品有"奇怪的酒店"、东京大学综合图书馆新馆等。

He is an architect, associate professor of the University of Tokyo and the director of Kawazoe Lab. He is the principal architect of kousou Inc. and the chairman of Japan and Netherlands Architecture Cultural Association. Key projects: "Henn-na Hotel", Annex Comprehensive Library, the University of Tokyo.

总建筑师
Chief Architect

章明
Zhang Ming

同济大学建筑与城市规划学院建筑系副主任、教授、博士生导师，同济大学建筑设计研究院（集团）有限公司原作设计工作室主持建筑师。

Professor / Vice Head of Department of Architecture, College of Architecture and Urban Planning, Tongji University; Chief Architect of TJAD Original Design Studio.

空间艺术版块团队
Main Members of Urban Space Art

薮田尚久、关口正洋、前田礼、柯思羽、龚珏、山本朔太郎、飞田洋二、林季阳、卢钰、郭宁、草野充子、三轮良惠、阿部雄介、Keith Jun OSU、井上尚子、增井辰一郎、松浦麻基

Naohisa Yabuta, Masahiro Sekiguchi, Rei Maeda, Ke Siyu, Gong Jue, Sakutaro Yamamoto, Yoji Hida, Lin Jiyang, Lu Yu, Guo Ning, Mitsuko Kusano, Yoshie Miwa, Yusuke Abe, Keith Jun OSU, Hisako Inoue, Tatsuichiro Masui, Maki Matsuura

规划建筑版块团队
Main Members of Architecture and Urban

张海翱、徐航、游猎、杜骞、张子琪、海伦·洛克海德、安斯利·玛瑞、乔治·乔尼庚、上海市城市规划设计研究院、汪灏

Zhang Haiao, Xu Hang, You Lie, Du Qian, Zhang Ziqi, Helen Lochhead, Ainslie Murray, Giorgio Gianighian, Shanghai Urban Planning and Design Research Institute, Wang Hao

总建筑师团队
Main Members of Architects

秦曙、张姿、张斌、刘宇扬、柳亦春、刘毓劼、单文慧、杨晓青、潘陈超、苏婷、李晶晶

Qin Shu, Zhang Zi, Zhang Bin, Liu Yuyang, Liu Yichun, Liu Yujie, Shan Wenhui, Yang Xiaoqing, Pan Chenchao, Su Ting, Li Jingjing

公共艺术作品制作协调团队
Survived Art Works of Production Coordination Team

欣稚锋艺术发展（上海）有限公司

Art Pioneer Studio

主视觉设计团队
Main Visual Design Team

韩家英设计有限公司

Han Jiaying Design Co., Ltd.

布展团队
Exhibition Executive Team

上海风语筑展示股份有限公司

Shanghai fengyuzhu Exhibition Co.,Ltd.

主题阐释——相遇
Encounter

上海是一座依水而生的城市，黄浦江、苏州河是上海的母亲河。一个世纪以来，"一江一河"见证了上海从一个小渔村到现代化大都市的成长，也见证了上海近代工商业的发展。浦江两岸曾经厂房、码头、仓库林立，很长一段时间以来，只有外滩是人们可以亲近浦江的地方，在其他的岸线段，人们听得到汽笛的鸣响却很难见到江水的波涛。

2010 年世博会开启了黄浦江岸线功能转型的序幕，也标志着上海的城市发展进入越来越注重城市品质建设的阶段，决心要用美好的空间来回应人民对美好生活的期盼。

2017 年 12 月，在市委、市政府的重要部署和沿线各单位持续不断的努力下，黄浦江两岸终于实现了 45 公里岸线贯通。这一公共空间建设的伟大壮举是近年来上海城市更新最重要的作品之一，不仅为上海市民提供了景观优美的活动空间，也以其深刻的人文关怀为世界城市建设史增添了亮丽的乐章。

2019 年 1 月 31 日，市政府进一步出台了《关于提升黄浦江、苏州河沿岸地区规划建设工作的指导意见》，批复了《黄浦江沿岸地区建设规划（2018—2035 年）》和《苏州河沿岸地区建设规划（2018—2035 年）》，这几个纲领性文件进一步明确将黄浦江规划定位为全球城市发展能级的集中展示区，将苏州河沿岸规划定位为特大城市宜居生活的典型示范区。

总结"一江一河"公共空间建设的既有成就，展望"一江一河"宏观战略的美好蓝图，2019 上海城市空间艺术季的主题方向应运而生。本届空间艺术季将通过艺术植入空间的活动，邀请人们亲身感受滨江贯通这一公共空间作品；将通过展览与实践相结合的方式，搭建一个对话的平台，邀请人们共同探讨"滨水空间为人类带来美好生活"这一世界性的话题。

主展主题由总策展人提案确定为"相遇"，意喻本届艺术季所呈现的人与人的相遇、水与岸的相遇、艺术与城市空间的相遇、历史与未来的相遇，将激发更多美好生活、美好情感的相遇，在庆祝新中国成立 70 周年之际，成为人们不可磨灭的城市记忆。李强书记指出，要把黄浦江两岸打造成城市的客厅。客厅已造就，2019 上海城市空间艺术季邀请全世界人民与上海相遇。

Shanghai is a city born by and of waters. Its twin mothers, Huangpu River and Suzhou Creek, have witnessed how Shanghai evolved from a small fishing village to a modern metropolitan in the course of a century, and how a modern industrial and business world has developed here. The banks of Huangpu River used to be a forest of plants, wharfs and warehouses, and for a long time the Bund was the only section along the river with intimate terms with people while wherever else the tides were overwhelmed by the steam whistling.

Thanks to the EXPO 2010, the functions of Huangpu banks began to change, which marked a new stage of Shanghai urban development characterized by increasingly emphasis on urban qualities and determined responses to the citizens' prospect of lovely life with beautiful spaces.

In December 2017, by virtue of the significant arrangements made by Shanghai Municipal Committee of CPC and Shanghai Municipal People's Government as well as the continuing efforts of all involved units, the 45km Huangpu Waterfront Connection, a great feat of public space and one of the most important results of recent urban renewal in Shanghai, was completed. The project offers scenic activity spaces for the citizens and adds a brilliant chapter to the global history of city construction with its deep humane concerns.

At 31 January 2019, Shanghai Municipal People's Government introduced the *Guiding Opinions on Promoting the Planning and Constructions Along Huangpu River and Suzhou Creek* and officially approved the *Construction Plan Along the Banks of Huangpu River (2018-2035)* and *Construction Plan Along the Banks of Suzhou Creek (2018-2035)*. These programmatic documents further confirm the target of Huangpu River banks as a concentrated exhibition area for the developmental capacity level of a global city, and Suzhou Creek banks as a typical demonstration area for a livable megacity.

With summarizing the achievements in constructing public spaces along the Huangpu River and the Suzhou Creek and visioning the marvelous future of the River and the Creek from a macro perspective, the theme of SUSAS 2019, which begins at the end of September, is thus determined. SUSAS 2019 invites people to personally feel the Huangpu River Connection as a public space by implanting art into spaces and to discuss the global issue "how waterfronts bring wonderful life to people" at the forum created by incorporating exhibitions with practices.

The theme of main exhibition, as proposed by the chief curator, is "encounter" for encounters of people, water and bank, art and urban spaces, history and future which in turn incite more beautiful life and emotional encounters and create inerasable city memories at the 70th Anniversary of the People's Republic of China. Li Qiang, Municipal Party Secretary of Shanghai, pointed out that the banks of Huangpu River should be the city's parlor. Now that the parlor is completed, SUSAS 2019 is inviting the world to encounter Shanghai.

策展思路
Thoughts of Curation

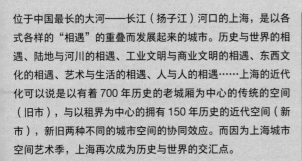

位于中国最长的大河——长江（扬子江）河口的上海，是以各式各样的"相遇"的重叠而发展起来的城市。历史与世界的相遇、陆地与河川的相遇、工业文明与商业文明的相遇、东西文化的相遇、艺术与生活的相遇、人与人的相遇……上海的近代化可以说是以有着 700 年历史的老城厢为中心的传统的空间（旧市），与以租界为中心的拥有 150 年历史的近代空间（新市），新旧两种不同的城市空间的协同效应。而因为上海城市空间艺术季，上海再次成为历史与世界的交汇点。

本届艺术季，以"相遇"为大主题，通过三个分主题——"百年工业与百年人文""源自河流的艺术""连接世界的网络"，进行演绎。通过对黄浦江沿岸的杨浦滨江港湾设施以及工业遗产以艺术的形式进行再生，提高上海城市空间的魅力。

"百年工业与百年人文"通过活化使用已经完成使命的船坞和船只等的工业遗产，从而围绕其展开艺术作品，映射 19 世纪下半叶至 20 世纪上半叶生活在国际都市上海的人们的生活文化和记忆的艺术作品。

"源自河流的艺术"通过与河流融为一体的艺术作品，展示从河道开始扩展的 16 个行政地区的独特魅力。有曾经是租界的各国风情的中心地区，也有老县城风韵犹存的地区，还有广阔的农田以及留存未经加工的自然的地区，多样的 16 区魅力呈现了"上海 16 景"。

一个城市魅力的根源在于步行所能体验到的趣味。"连接世界的网络"，通过结合上海的文化和艺术，创造出世界一流的城市体验。人们在漫游滨江栈道时，将会一个接一个地与艺术品在风景中相遇，这无疑将是一次难忘的经历。

Shanghai, located at the mouth of the Yangtze River, the longest river in China, is a city that has developed through the overlap of various "encounters": the encounter between history and the world, land and river, industrial and commercial civilizations, the East and the West, art and life, and between different people, etc.. The modernization of Shanghai can be described as the synergy of the 700-year-old Old City area, and the 150-year-old modern area around the concession. And Shanghai has once again become a meeting point of history and the world due to the Shanghai Urban Space Art Season.

This year's art season takes "encounter" as the main theme and sets three sub-themes: "a century of industry and humanities", "art from the river" and "network connecting the world". It is committed to enhancing the charm of Shanghai's urban space through the artistic regeneration of the Yangpu Waterfront harbor facilities and industrial heritage.

"A century of industry and humanities" revitalizes and makes use of the industrial heritage such as docks and ships to create artworks. These works reflect the life and memory of people living in Shanghai during the late 19th century and the early 20th century.

"Art from the river" is a section of artworks that is integrated with the river, displaying the unique charm of the 16 administrative regions that have expanded from the river. It shows the diverse charm of different districts through "16 Sceneries of Shanghai", including central areas that were once the concessions, places remaining the charm of old towns, and vast farmlands and areas where unprocessed nature remains.

The fundamental charm of a city lies in the fun that can be experienced while walking around. "Network connecting the world" creates a world-class urban experience by combining the culture and art of Shanghai. People can see artworks in the landscape as they walking along the riverfront, and it will undoubtedly be an unforgettable experience for them.

基于城市文脉、激发城市活力的艺术展
An Art Exhibition Based on Urban Culture and to Stimulate Urban Vitality

北川富朗
2019 上海城市空间艺术季主展览总策展人

Fram Kitagawa
Artistic Director of SUSAS 2019

我在日本作为艺术策展人，通过策划各式各样的艺术活动来为城市或乡村增添活力。比如，越后妻有大地艺术节是在大雪地带的山区里，濑户内国际艺术节是在人口稀疏的海岛上，等等。这对今天的社会来说，就像是把人与土地，把人与人联系在一起的一种尝试。

或许因为这些策展经历，上海城市空间艺术季的主办方就邀请我来担任 2019 年的策展人。这次来上海工作，我很开心，因为我对上海的历史，以及中日两国自古以来的交流非常感兴趣。虽然中日两国各自具有不同的文化体系，但日本从古代开始就向中国学习，可以说构成了日本文化的骨架。这次能参与中国的项目，为此贡献一点力量，对我来说是非常幸运的。

杨浦滨江初印象

大约在 2017 年，我第一次来到杨浦滨江，那时候很多都还不确定，到了 2018 年开始着手策划，年末就决定以杨浦滨江作为展览的场地。刚来到杨浦滨江的时候，感觉既新鲜又感动，觉得这条河真叫人喜欢，特别是当置身于岸边，看到河面上来来往往的运送货物的船，那种生机勃勃的、散发着生活气息的场景，给我留下了深刻的印象。

兼容并蓄的"相遇"

上海是一个有故事的地方，从很久以前就开始建立城市，也有很多长居于此的居民。有趣的是，上海也非常国际化，作为港口，在不同的时代里都与其他国家有许多交流，一直是中国对外的窗口。过去和现在，东方和西方，不同的文化在这里产生精彩的相遇。

2019 年的参展艺术家也是如此，不仅有中国本土的艺术家，还有来自世界各地的艺术家，比如，瑞士的菲利斯·瓦里尼、英国的理查德·威尔森，等等。他们和他们的作品在这里相遇，是一件非常有趣的事。

"相遇"这个主题是对上海的既往及其发展过程的思考，并对各种各样的新的相遇怀有期待。我也希望以上海为起点，参与到更多与中国的丰富的社会生活有关的项目中去。

与城市紧密关联

2019 上海城市空间艺术季具有区别于其他艺术展的独特之处——与城市紧密关联。首先，艺术季与杨浦滨江 5.5 公里的

As an art curator in Japan, I plan various art activities to add vitality to the city or the country. For example, the Echigo-Tsumari Art Triennale is in the mountains in the snowy area, the Setouchi Triennale is on the sparsely-populated island. In today's society, this is like an attempt to connect different people and people to land.

Perhaps because of my curatorial experiences, I was invited to be the curator of the 2019 Shanghai Urban Space Art Season. I am very happy to work in Shanghai this time because I am very interested in the history of Shanghai and the exchanges between China and Japan since ancient times. Although the two countries have different cultural systems, Japan has been learning from China since ancient times which formed the backbone of Japanese culture. It is very fortunate for me to be able to participate in and contribute to such a Chinese project this time.

My First Impression of Yangpu Riverfront

I first went to the Yangpu Riverfront in about 2017. There were many uncertainties at that time. I started planning in 2018 and decided to use Yangpu Riverfront as the venue for the exhibition at the end of the year. When I first came to Yangpu Riverfront, I felt both fresh and moved. I was so into this appealing river, especially when I was on the shore and looking at the cargoes coming and going on the river. Such a vibrant, life-like scene has left a deep impression on me.

An Inclusive "Encounter"

Shanghai is a place full of stories and has many inhabitants who have been living here since a long time ago. Interestingly, Shanghai is also very international. As a port city, Shanghai exchanges a lot with other countries at different times and has always been a window to the outside world. Encounters between the past and the present, the East and the West, and different cultures happen in this city.

The same is true for the participating artists in 2019. There are not only local Chinese artists but also artists from all over the world, such as Felice Varini from Switzerland and Richard Wilson from the United Kingdom, etc.. Interestingly, these people and their works can meet here.

The theme of "encounter" is a reflection on Shanghai's past and its development process, and contains expectations for various new encounters. I also hope to take Shanghai as a starting point for more projects related to China's rich social life.

Closely Connected to the City

The 2019 Shanghai Urban Space Art Season has a unique feature that sets it apart from other art exhibitions - its close connection to the city. First of all, the exhibition is integrated with the development of the 5.5 km Yangpu Riverfront. Outstanding artists from all over the world gather here to create different works at 20 points along the riverfront, and their artworks will be preserved there. Secondly, by thinking about what kind of city Shanghai is, 16 Chinese artists are invited to capture the characteristics of Shanghai through their own perspectives and present them in the historic Maoma warehouse, in order to show the

开发工程相结合：来自世界各地的优秀艺术家汇聚于此，在滨江空间的 20 个点位上进行不同的创作，并且其艺术作品会长久地保留在这里。其次，以"上海到底是怎样一座城市？"为出发点，邀请 16 位中国艺术家，通过各自不同的视角捕捉上海的特色，并在历史遗存建筑毛麻仓库中呈现，从过去的上海的划分到现在的"大上海"概念，展现上海的开放性和包容性。

关于艺术家的选择

2019 年的艺术季有来自世界各地的具有代表性的艺术家，以及许多不同类型的作品。我对于艺术家的选择，总的来说是从三个方面来考虑的。

首先，2019 年的室外展场是一段相对狭长的区域，空间本身的变化就非常有趣，但仍然需要一些装饰性的内容，所以邀请不同类型的艺术家在这里创作，能给行走其中的人们一个又一个不同的新鲜体验。

其次，艺术家们进入杨浦滨江时，会遇到各种各样的事物，他们要能够与这个地方产生联结。比如这里有工厂，有劳动者，早上附近的居民还会聚集在一起进行各种活动。艺术家需要先去理解这些只属于这个区域的东西，然后把这里的人和他们的生活在这块场地上重新应用和呈现出来。其中最重要的是，所有的作品、活动都欢迎公众参与进来，通过他们的行动、言语，以及他们之间的关系，去展现这个地区的特征，告诉人们这里究竟是一个怎样的地方。

最后，根据黄浦江的区域特性来选取与之相适合的艺术家。黄浦江两岸在数百年间经历了经济、文化等方面的蓬勃发展。不久之前这里还是工业区，与国外往来运输大量的货物，也与中国内陆地区保持紧密的联系。人与物在这里汇集，住宅、学校等城市元素丰富多样。以河流为中心来思考城市规划是十分明智的，假如只以道路为中心，比如不断建造高楼大厦，就会给人带来压迫感，而河流可以形成一个非常舒展的空间，容纳各种可能的公共活动。这对普通市民来说非常重要。

让艺术走出美术馆

美术馆或者画廊里的作品，只有某些固定的观众群体会去参观，比如想要从事艺术创作的，或者正在学习艺术的人。因此，所谓的公共空间是指那些与艺术无关的人所处的场所。不管人们是喜欢还是讨厌，开心还是难过，把他们所持有的种种情感、情绪连在一起，就是公共艺术的作用之一。可以说，这样的艺术，并不是被保存起来受人观赏的，而是要在社会中与各种各样的人和事相遇。换句话说，世界上的城市各有精彩之处，在这些城市中行走时，如果公共艺术使人们产生一种真实的、切身的对于这个城市的独特感受，认识到与自己不一样的观念和思考方式，那就是"相遇"。在我看来，这是公共艺术的真正意义与价值所在。

2019 年的艺术季还针对当地的市民，特别是儿童，开展了艺术工作坊等活动，试图以更多样的方式让公众参与到其中。因此本届艺术季可以说是一次契机，让普通的市民有机会去思考上海这座城市的未来，或者说，从现在起这座城市该如何发展。这正是我所期待的。

city's openness and inclusiveness from the old city division in the past to the concept of "Greater Shanghai" in the present.

About the Artists

The 2019 Art Season has representative artists from all over the world, as well as many different types of works. My choice of artists is generally considered from three aspects.

First of all, the outdoor exhibition area this year is relatively long and narrow. The changes in the space itself are very interesting, but we still need some decorative content for space. Therefore, different types of artists are invited to create here, which can give people a different and fresh experience while walking in the space.

Secondly, the artists need to be able to connect with the place when they enter the Yangpu Riverfront space. For example, there are factories and laborers here. Every morning, people living in the neighborhood will gather for various activities. The artists need to first understand these things that only belong to this area, and then reapply and re-present the people and their lives on this site. One of the most important things is that we welcome the public to participate in all of the works and activities, to show the characteristics of this area through their actions, words, and their relationship with each other, and to tell people what kind of place it really is.

Finally, the artists are selected to match the regional characteristics of the Huangpu River. Both sides of the Huangpu River have experienced vigorous economic and cultural development over the centuries. Not long ago it was still an industrial area, transporting large amounts of goods to and from foreign countries while maintaining strong ties with the interior of China. People and things came together here, and there was a rich diversity of urban elements such as houses and schools. It is wise to make urban plannings with the river at its core, as it would be oppressive if it were centred only on roads. The constant construction of tall buildings can depress people living in the city. However, the river can create a very extended space for all possible public activities. This is very important for ordinary citizens.

Bring the Art Out of the Gallery

The works in art museums or galleries only have certain fixed groups of audiences, such as those who want to create artworks, or those who are studying art. The so-called public space is a place where people who have nothing to do with art live. Whether people like it or not, are happy or sad, connecting their various feelings and emotions is one of the roles of public art. This kind of art, so to speak, is not preserved for watching, but is to meet with people and events. In other words, cities in the world have their own features. If public art can create a real, personal and unique feeling about the city as one walks through it, and make one realizing these different ideas and ways of thinking, we can call it an "encounter". In my opinion, this is the true meaning and value of public art.

The 2019 SUSAS also organizes art workshops for local citizens, especially children, in an attempt to engage the public in a more diverse way. So this session can be considered as an opportunity for ordinary citizens to think about the future of Shanghai, or rather, how the city should develop from now on. That's exactly what I'm looking forward to.

水·美好生活·策展——2019 上海城市空间艺术季规划建筑版块反思[1]

Water & Life & Curation — Reflections on the Planning and Architecture Section of the 2019 Shanghai Urban Space Art Season[1]

阮昕

2019 上海城市空间艺术季规划建筑版块策展人

Ruan Xin

Architecture and Urban Curator in SUSAS 2019

理论基础

作为一个业余策展人，代表上海交通大学设计学院，担任如此重任，我考虑这个展览要有一定的学术性。于是我们试图为"滨水空间创造美好生活"这个主题做一个理论诠释。首先，"美"和"好"不可分割。苏格拉底式的"美好"与中国儒家思想对于美德的定义不谋而合，都为美好生活赋予了道德的分量。这既非单纯的享乐主义式的幸福生活，亦非现代人所谓生活方式的选择。

其次是水的象征含义：在我们展示城市建筑依水而生，为人的生活带来愉悦的背后，水所象征的纯净性与永续再生之恒力，是一种不断驱动着城市与生活转变与再生的魔力。

策展即是讲故事

上述的理论框架自然不能以教化方式在展览中呈现，否则连"打卡"的年轻人都不去了，何谈公众参与？作家西蒙·莱斯说过，历史学家其实是描写过去的小说家（他也接着说，小说家即是描写现实的历史学家……）。于是我们非常历史性地选择了"三城记"的滨江空间故事——上海、威尼斯和悉尼。当然如此选择必定也有随机性。悉尼入选与我在这个城市生活了 20 多年多少有点关联。而选择这三个城市的最主要原因是，我们今天去旅游打卡，观其象却并不一定了解或会去深思滨水空间的故事与这三个城市精气神的关系。

上海：我们今天讲上海的滨水空间，往往只讲"一江一河"。可是上海的"性格"，她的方言，她的精明与她的包容，如何不是江南水乡的"场地精神"呢？千年来，上海这么一个偏远

Theoretical Basis

As an amateur curator representing the School of Design of Shanghai Jiaotong University to shoulder such an important task, I consider that this exhibition should have a certain academic nature. So we tried to make a theoretical interpretation for the theme of "waterfront space brings a better life of mankind". First of all, the "beauty" and "goodness" of life are inseparable. The Socratic concept of "good" coincides with the Chinese Confucian definition of virtue, both weighing morality a lot. This is neither a pure hedonistic happy life nor is it a so-called lifestyle choice for modern people.

Secondly, think about the symbolic meaning of water. We display the urban architectures alongside the water, bringing pleasure to people's lives. More profoundly, the purity and constant power of water regeneration are like magic that constantly drives the transformation and regeneration of city and life.

Curation is to Tell the Stories

The above theoretical framework certainly cannot be presented rigidly in the exhibition. Otherwise, even the young people who are keen on fancy new things will not willing to go there, how can we realize public participation? The writer Simon Leys said that historians are in fact novelists writing about history, and he also said that novelists are historians telling about reality...... So we chose the tale of the riverfront space of three cities - Shanghai, Venice, and Sydney. Of course, there must be some randomness in this choice. The selection of Sydney is somewhat related to the fact that I have lived in the city for more than 20 years. The main reason for choosing these three cities is that when we visit them today, we do not really understand or think about the relationship between the history of waterfront space and the essence of the city.

When we talk about Shanghai's waterfront space today, we often refer to the Huangpu River and Suzhou Creek. But Shanghai's "character", dialect, people's smartness, and tolerance are also part of the soul of a water town. Thousands of years ago, Shanghai was a small remote water town with dense river channels and embankments and surrounding rice farming fields. Our ancestors have already started to use the circulating tides of the Huangpu River to regulate irrigation, water supply, and sewage. Such an "ecological water town" was severely destroyed by the opening of the port. The Europeans built roads and villas in the concession area, filling in large and small waterways and discharging waste

1 本文原载于《时代建筑》2020 年第 1 期，总第 171 期。

1 This article was originally published in *Time Architecture*, 2020, (1), 171.

的小市镇，其实是一个独特的水乡。密集的河道与堤圩造就了环绕的稻米农耕田地。先民们早已开始利用黄浦江神奇的循环潮汐来调节灌溉、供水与排污。如此"生态水乡"，持续到开埠后，其"生命血脉"才遭受严重堵塞。欧洲人在租界筑路并搭建洋楼别墅，大大小小的水道被填埋，废物通过下水管道排放到中国贫民难民聚居的开放河道，曾经如同"血脉"般连接黄浦江的水道被逐渐切断。剩下苏州河与黄浦江，倒是为中国几乎最早的现代工业城市提供了交通便利。

今天上海"一江一河"的故事，是一个不断恢复水乡城市生态的举措，因为上海不应该，也不会失去滨水品质。只有重建百川，方可做到真正的"海纳"！

威尼斯：这个如同童话般的滨水城市始于一个庇护所。罗马帝国的败落，令北方的蛮族将避难者们驱赶到了这片沼泽地寻求庇护。起初，海水为未来的威尼斯人提供了防御与安全，这是因水而造成的隔离。但这是一个没有淡水水源的孤岛，正因为缺水，才给威尼斯的工匠们带来城市建筑的技术挑战。

水路与蓄水使这个偏僻的难民营迎来了新生：通过水上贸易，共和国创造了财富与繁荣；以奇迹般的水路策划，获取圣人舍利，水为共和国提供了宗教合法性；蓄淡水、防海水为城市形式、建筑和艺术生活提供了无尽灵感源泉。正是水，令威尼斯成为了世界最精美的城市艺术品之一。

悉尼：1787 年 5 月 13 日，拥有 11 支帆船的第一批舰队从英格兰抵达悉尼湾，建立起了一个流放殖民地。这就是澳大利亚并非吉利的开端。与威尼斯不尽相同，对于英格兰的立法者而言，水不仅创造了一个不可逾越的监牢以安置源源不断的犯人，并且还要成为重塑他们品行的良方。不知是否超出了英国立法者的想象，今天的澳大利亚人成为了守法的公民；悉尼港从承接罪犯的登陆点，到为后来工业提供生命线，而现在摇身一变成为全世界最美的滨水休闲港，公平而开放地面向其多元文化的居民。

展呈

故事如何通过展览的方式来讲述？如何在呈现自然元素、技术和形式建造的同时，突显赋予物之灵性的人。换言之，这个展览是通过"水之魔力"颂扬"人的魔力"。我们不妨仍以威尼斯和悉尼为例吧。

威尼斯主题展在形式上简单而优雅，目的是要追求"意馀于形"，以便将其内涵——即威尼斯的建造史、文化史，甚至里面的人物与事件，娓娓道来。观者若潜心"读"进去，就会学到东西。

说到人物，威尼斯展中就有这么一位，我要求威尼斯团队务必精心描述。我以前时常去威尼斯，也在威尼斯建筑大学教过暑期课程，所以认识了当地的雷纳托·里齐教授。里齐教授（也就是近年来快退休了才提到教授。全世界又何尝不是弄潮儿更

through sewer pipes into the open river where the Chinese poor refugees lived, gradually cutting off the waterways that were once connecting the Huangpu River, leaving the Suzhou Creek and Huangpu River to provide convenience for transportation in this almost the earliest modern industrial city in China.

The construction of the Huangpu River and Suzhou Creek in Shanghai today is an initiative to continuously restore the urban ecology of a water town. Only by rebuilding rivers can the real "sea" be achieved.

Venice, a fairy-tale waterfront city, was a sanctuary at the beginning. The Roman Empire fell, and the barbarians from the north drove the refugees to this marshland. At first, the sea provided defense and security for the people, the later Venetians, by creating isolation with the water. But this was an isolated island without freshwater, and it was the lack of freshwater that presented Venetian artisans with the technical challenges of urban construction.

Waterways and water storage brought new life to this isolated "refugee camp". The Venetian Republic created wealth and prosperity through water-borne trade. Venetians obtain the relics of saints with waterway planning, legitimating the country's religion. The country took the strategy of storing fresh water and protecting from seawater, giving an endless source of inspiration for urban formality, architecture, and artistic life. It is water that has made Venice one of the world's most exquisite urban works of art.

On May 13, 1787, the first fleet of eleven sailing ships arrived in Sydney Cove from England to establish an exile colony. This was the beginning of Australia, a beginning that was not auspicious. Unlike Venice, for England's legislators, water not only created an impenetrable prison to house the prisoners but also became a good thing for reshaping their character. No one knows if it was beyond the imagination of the English legislators that Australians today are law-abiding citizens. Sydney Cove, being a landing point for convicts and supporting later industry, is now transformed into the world's most beautiful waterfront recreation harbor, fair and open to its multicultural inhabitants.

Presentation

How to tell the story through exhibitions? How to present the natural elements, techniques, and construction forms while highlighting the people who give the things spirituality? In other words, this exhibition should manifest the "power of mankind" through the "power of water". Let's take Venice and Sydney as examples.

The Venice theme exhibition is simple and elegant in form, to pursue "meaning beyond the form" in order to express its connotations, namely, the history of Venice's construction and culture, and its people and events. If the viewer concentrates on "reading" in, he will learn something.

Speaking of people, there is someone in the Venice exhibition deserving to be addressed carefully. I used to go to Venice often and I once taught summer courses at the Venice University of Architecture, so I knew a local professor called Renato Rizzi. Professor Rizzi is a great ordinary architect. But the fact is that he hasn't been promoted to professor title until recent years when he was about to retire. He doesn't chase fame and wealth. Truly being an artist, he values what he does more than himself. One of the things he has been dedicated to is making models out of plaster. He believes that after repeated work, the plaster will eventually have spirits as if it will talk to you. You can get a glimpse of his works in the Venice exhibition. He spent 25 years designing and building a theater in Poland, which is a high-quality work. This theme exhibition uses a section to introduce to the audience an ordinary professor, an aspiring artist, and an architect. Venice makes him who he is.

容易功成名就呢？）是一个了不起的普通建筑师。他不追逐名利，但把自己做的事看得比他个人更重，这是一种真正的艺术家品质。他最潜心做的一件事情就是用石膏做模型，认为经过反复的劳作，石膏最终也会有灵性，好像会跟你说话。大家在威尼斯展中可窥见一番。他花了25年的时间在波兰设计建造了一间戏院，是一个非常有品质的作品。这次主题展将这么一个因滨水威尼斯而造就的人物做成了一个版块，让观众认识了解一位普普通通的教授，一个有追求的艺术家，一个建筑师。

悉尼主题展跟威尼斯的就完全不一样了。这是个非常有氛围的围合空间。展览把悉尼的港湾分割成很多块，用非常精细的方式描绘出来，以装置展示。三十多公里的海岸线有很多的内容，我第二次去看的时候，发现甚至能找到我在悉尼的家。

对于悉尼海湾的品质，它的微妙性、气氛与植被，甚至是否能感受到它的光线与色彩，想象中闻到桉树飘出来的清香，我觉得这都是需要有点心境去体会的。这种细致的程度和耐看性，是我们展览的一点追求，追求一种氛围，一种内涵。

The Sydney theme exhibition is completely different from the Venice one. This is an enclosed space with nice vibes. The exhibition divides Sydney's harbor into many sections, depicted in a very detailed way and displayed in installations. There is so much of the display of the 30-kilometer coastline that when I went there for the second time, I could even find my home in Sydney.

As for the exhibition of Sydney Cove, I believe if you are mindful enough, you can feel its subtlety, atmosphere, vegetation, light and colour, and smell the fragrance wafting out of the eucalyptus trees. This degree of meticulousness and endurance is a bit of the pursuit of our exhibition, the pursuit of an atmosphere and a connotation.

从历史原真的叠合出发，回归人与城市的关系——百年船坞与杨浦滨江的改造设计

Returning to the Relationship between Man and the City Based on Overlapping History—Reconstruction Design of the Shipyard and Yangpu Waterfront

章明
2019 上海城市空间艺术季总建筑师

Zhang Ming
Chief Architect of SUSAS 2019

2019 上海城市空间艺术季的主展场片区选在黄浦江西岸的船坞区域。船坞位于杨树浦路 468 号，隶属上海船厂浦西分厂，船厂内留存的并列双船坞空间形式，在黄浦江沿线中是绝无仅有的。其历史跨越百年，早在 1900 年，德资的瑞镕船厂就在这里开挖船坞，新中国成立后并入上海船舶修造厂，即如今的上海船厂。

工业空间的非日常体验

在最初进行选址勘察时，我们从小楼梯下到船坞深处 -10m 标高的地方，感到一种强烈的震撼。高差上的变化、工业空间的超常规尺度……都给人一种非日常的空间体验。船坞周围遗留的造船、修船痕迹，以及各种相关的大型工业构件也带来视觉上的美学冲击。

船坞空间的界面是独特的钢板桩墙面，表面上布满斑驳的痕迹。在漫长的时间里，修复船只时反复喷涂油漆的过程，被层层叠加，呈现在墙面的颜色之中。当油漆经风化部分剥落后，这种层叠的颜色呈现出来：墙体的外层可能为蓝色，里面一层是红色，再往里一层又是黄色……这反映了一种历史的叠合过程。从过去到现在，历史过程中发生的故事会带给人们想象，也使时间产生厚度。我们称之为"叠合的原真"，包括作为物质的斑驳墙面的叠合，以及作为抽象记忆的时间片段的叠合。

人站在船坞内，面向黄浦江的方向，尽头是高大的坞门，偶尔看到门上方有一根桅杆缓缓驶过，仿佛听见现实与历史的共振。

使船坞成为独特的空间印记

在主展场片区，船坞这个独具特色的空间，旨在形成一个令人难忘的空间印记，并与艺术作品发生互动。

The main exhibition area of the 2019 Shanghai Urban Space Art Season is located in the dockyard on the west bank of the Huangpu River at No.468 Yangshupu Road. The dockyard is part of the Puxi Branch of the Shanghai Shipyard and is unique along the Huangpu River for its parallel double-dock space. Its history spans over a hundred years. As early as 1900, the German-owned Ruirong Shipyard dug a dock here. After the founding of the People's Republic of China, it was incorporated into the Shanghai Ship Repair Yard, now the Shanghai Shipyard.

Unusual Experience of Industrial Space

During the initial site survey, we were strongly shocked when we descended from the small staircase to the depth of the dockyard at an elevation of -10m. The difference in height and the extraordinary scale of the industrial space gave us an unusual space experience. The traces of shipbuilding and repairing left around the dockyard and various related large industrial components also brought us great visual aesthetic impact.

The interface of the dockyard is a special wall of steel sheet pile with its surface covered with mottled marks. During its long history, the wall has been repeatedly painted for ship restoration. When the paint is partially peeled off by weathering, the painting colours under the surface now become visible: the outer layer of the wall may be blue, the inner layer is red, and the layer further in is yellow ... This reflects a kind of overlapping history of different periods. From the past to the present, history always makes people imagine a lot and gives the passage of time meaning. It is called the "overlapping reality", which includes the overlapping of the mottled wall as material and the time fragments as abstract memories.

Stand in the dock facing the direction of Huangpu River. The end is the tall dock gate. Occasionally above the door, a mast slowly passes by. It seems like the reality is resonating with history.

Make the Dockyard a Particular Space Mark

In the main exhibition area, the dock is designed to form an unforgettable space mark and interact with artworks.

There are two docks, one large and one small. The small dock is used as a venue for the opening ceremony and for hosting multimedia artworks during the art season. In the

船坞共有两个，一大一小。小船坞被用作开幕式的场地，同时在艺术季期间承载多媒体艺术作品。到了晚上，在艺术品的灯光配合下，小船坞内流光溢彩，仿佛与艺术作品共同构成一件大型装置。考虑到人流的上下和疏散，我们使用 6300 根脚手架杆件进行连接，搭建了一个大的台地，当中还有一个船形舞台。这种脚手架的搭建是一种装配式的建造方法，能够快速搭建，未来也能快速拆除。

大船坞内新建了一个可容纳 500 人的报告厅，是一个能遮风避雨的相对封闭的空间。屋顶兼具看台的功能，采用钢木结构，保证人流能从两侧通过。大船坞考虑在艺术季期间，以及展览过后融入日常的使用可能，使得空间可以兼容多样的活动，包括论坛、时装秀等，以此将原来的工业空间转化为当下能够使用的、具有艺术气质和时尚气质的空间。

我们在入口处还用脚手架搭建了一个安检棚，包含一定象征性，也解决了取票、志愿服务、物品寄存等功能，满足大型展览的需要。同时，安检棚作为入口大门与船坞内的台地、报告厅，以及毛麻仓库等，形成完整的空间序列，以装置艺术作品的方式呈现在观众面前。

艺术品与所在场所相锚固

2019 上海城市空间艺术季的特别之处在于，艺术品不是完全陈列在艺术馆内，而是试图跟其所在的场所相锚固。它们既像是从场所中生长出来，又游离在场所之上。对我来说，总建筑师的工作之一是协助总策展人和艺术家尽可能地了解场地条件，创作在地作品。我希望作品与空间能够融合，艺术家与建筑师所营造的不同类型的空间能够真正"相遇"。

另一方面，要对主展场的杨浦滨江南段 5.5 公里滨水公共空间进行系统化梳理。其中包括三个主要的场馆，或者说是驿站。船坞区域由毛麻仓库和船坞构成主展场片区；中段的驿站是由烟草仓库改造的"绿之丘"承担；在收尾的东段，杨树浦发电厂外的灰仓被用作第三个驿站。除此之外，入口的布置、配套服务设施的设计、动态交通和静态交通的梳理等，也是建筑师的工作之一。

三种"相遇"

相遇，首先是艺术和公共空间的相遇。将艺术作品沿城市空间分布，使其与空间产生良性的互动，二者有机地融为一体。这样，不仅可以提升公共空间的品质和艺术感，同时，公共空间也为艺术品的存在提供了合理性。

另一方面，杨浦滨江具有厚重的工业感和历史感，所以相遇也是历史与现在的相遇。这种相遇要有故事性，使空间发人深思。像原来厂区里的老工人所说，"在这里（改造后的杨浦滨江）感觉自己的过去跟现在连在一起"。

evening, the small dock is illuminated by the artworks, as if they were a large installation together. Considering the flow of people and evacuation convenience, we used 6,300 scaffolding poles to connect and build a large platform with a boat-shaped stage in the middle. This scaffolding is built as an assembly method of construction, allowing for quick erection and dismantling.

A new lecture hall with a capacity of 500 people has been built in the large dock, which is a relatively closed space that can shelter from wind and rain. The roof, which functions as a grandstand, is constructed of steel and wood to ensure that the flow of people can pass through from both sides. Considering the possibility of making the space available for daily use after the exhibition, the large dock is designed to be compatible with a variety of activities, including forums, fashion shows, etc., in order to transform the original industrial space into an artistic and artistic space with artistic and fashionable qualities that can be used today.

We also built a security shed at the entrance, which contains a certain symbolism and also functions as a place for ticket collection, volunteer service, and storage of goods to meet the needs of large exhibitions. At the same time, the security check shed, as the entrance gate, forms a complete spatial sequence with the terrace and the lecture hall of the dock as well as the Maoma warehouse, which is presented to the audience as an installation artwork.

Anchorage of the Art Work at Setting Spot

What is special about the 2019 Shanghai Urban Space Art Season is that the artworks are not completely displayed in the art galleries, but rather try to anchor themselves to the places they are located. They both seem to grow out of the place and stray from it. For me, part of my job is to assist the chief curator and the artist to understand the site conditions as much as possible and create works in situ. I hope that works and space can be integrated and that different types of spaces created by artists and architects can truly "encounter".

Besides, I need to systematize the 5.5 km waterfront public space in the southern section of the Yangpu riverfront, which is the main exhibition site. It includes three main venues, or let's say stations. The main exhibition area is composed of the Maoma warehouse and the docks; in the middle section, the station is the "Green Hill", which is a converted tobacco warehouse; in the eastern section, the gray warehouse outside the Yangshupu power plant is used as the third station. In addition, the layout of the entrance, the design of supporting services, and the arrangement of dynamic and static traffic were also part of the architects' work.

Three "Encounters"

First of all, the encounter between art and public space. Distribute artworks along the urban public space, making them well interact with the space while the two organically merging into one. In this way, the artworks can enhance the quality and artistic sense of the public space, and the public space also provides an area befitting those artworks.

Secondly, the encounter between the past and the present. Yangpu Waterfront has a long history of industrial development. We are supposed to tell the story through space and make it thought-provoking. As the old workers of the original Shipyard said, "Here (in the renovated Yangpu Riverfront), I feel that my past is connected to the present".

Last but not least, most importantly, the encounter between people. In the past, there were factories lined up south of Yangshupu Road, with high fences and warning signs saying "Staff Only". People living along the river even did not have access to the waterfront. These factories were once an important area of energy supply for Shanghai but have been gradually abandoned as the city has developed. It is becoming increasingly important to return the waterfront space to the citizens and make it a freely accessible urban shared space. The construction of the 45-kilometer riverfront space on both sides of the Huangpu River is a typical effort. In such high-quality waterfront spaces, more stories can really

此外，最重要的一个方面，是人与人的相遇。过去杨树浦路以南是鳞次栉比的工厂，到处都是高高的围墙，挂着"闲人免入"的警示牌，市民即便住在江边，也没有机会到达滨水空间。这些工业区虽然曾是上海的重要的能源供应地，但是随着城市和社会的发展，逐渐荒废。将滨水空间还给市民，使其成为可以自由到达的城市共享空间，变得越来越重要。上海黄浦江两岸45公里的贯通工程，正是还江于民的典型。在这样高品质的滨水空间中，通过人和人的相遇，可以真正地发生故事，诞生更多的历史。如今，每天黄昏时分，附近的市民摩肩接踵地来到这里纳凉、散步、健身、交流，空间与人的生活紧密地结合起来。这里不仅仅是一个观光地，更是百姓乐于日常使用的生活空间，这样的相遇是最有价值的。

happen through the encounter between people. Nowadays, every day at dusk, people from the neighborhood come here to walk, exercise and chat. The space is closely integrated with people's lives. It is not only a place for sightseeing but also a space that people are living in every day. The most valuable encounter always happens between people.

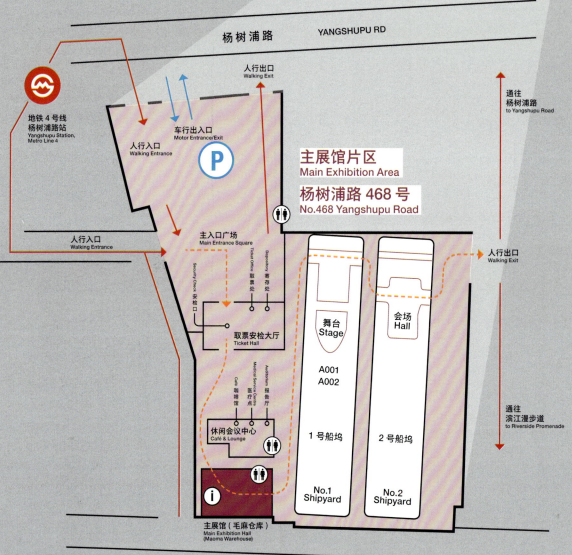

图例 Legend

- 码头渡口 Dock and Ferry
- 公交站 135 路、28 路 Bus Stop No.135 No.28
- 地铁 4 号线 Metro Line 4
- 主入口 Main Entrance
- 人行路线 Walking Lane
- 滨江步行线 Riverside Promenade
- 车行路线 Motor way
- 参观浏览路线 Tour Path
- 服务中心 Information Centre
- 停车场 Parking
- 卫生间 Toilet
- 主展馆片区 Main Exhibition Area
- 01 公共艺术作品 Public Artworks
- 计划中公共艺术作品 Public Artworks in Planning

地铁 4 号线 杨树浦路站 Yangshupu Station, Metro Line 4

秦皇岛路渡口 (其秦线) Qinhuangdao Road Ferry (to Qichangzhan Ferry)

主展馆片区 Main Exhibition Area 杨树浦路 468 号 No.468 Yangshupu Road

杨树浦水厂 Yangtszepoo Waterworks

THE HUANGPU RIVER

杨 树 浦 路 YANGSHUPU RD

人行出口 Walking Exit

车行出入口 Motor Entrance/Exit

通往 杨树浦路 to Yangshupu Road

地铁 4 号线 杨树浦路站 Yangshupu Station, Metro Line 4

人行入口 Walking Entrance

主展馆片区 Main Exhibition Area 杨树浦路 468 号 No.468 Yangshupu Road

人行入口 Walking Entrance

主入口广场 Main Entrance Square

人行出口 Walking Exit

安检口 Security Check

取票处 Ticket Office

寄存处 Depository

舞台 Stage

会场 Hall

取票安检大厅 Ticket Hall

A001 A002

咖啡馆 Café

医疗点 Medical Service Centre

报告厅 Auditorium

1 号船坞

2 号船坞

通往 滨江漫步道 to Riverside Promenade

休闲会议中心 Café & Lounge

No.1 Shipyard

No.2 Shipyard

主展馆 (毛麻仓库) Main Exhibition Hall (Maoma Warehouse)

秦皇岛路渡口 (其秦线) Qinhuangdao Road Ferry (to Qichangzhan Ferry)

黄 浦 江 THE HUANGPU RIVER

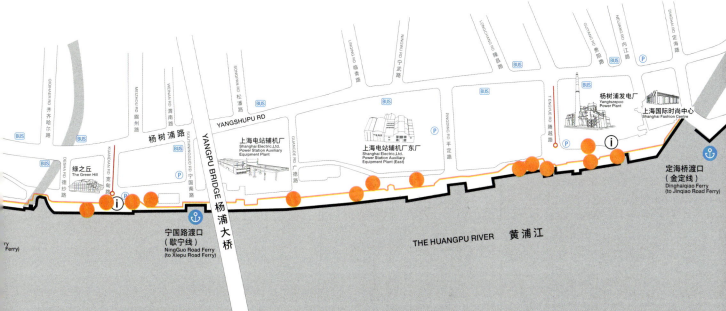

公共艺术作品
Public Artworks

杨浦滨江
（秦皇岛路码头—上海国际时尚中心）沿线
Yangpu Waterfront Huangpu River

空间艺术版块
Urban Space Art

A

A0 船坞记忆 Memory of Shipyard	1 号船坞 No.1 Shipyard
A1 水之相遇 Encounter Water	毛麻仓库 1F Maoma Warehouse 1F
A2 上海 16 景 16 Sceneries of Shanghai	毛麻仓库 2F Maoma Warehouse 2F

2019 上海城市空间艺
术季主展览 / 相遇
**2019 SUSAS Main
Exhibition / Encounter**

规划建筑版块
Planning & Architecture

P

P1 流水的心动：
黄浦江遇见悉尼
Heartbeat: Sydney Harbour
Juxtaposed with
the Huangpu River

P5 萍水相逢：
苏州河邂逅威尼斯
Serendipity of Water:
Venice Juxtaposed
with Suzhou Creek

P2 大气谦和
Magnanimity

P6 徜徉
Roaming

P3 善恶之水
The Virtue and Sin of Water

P7 乡愁
Nostalgia

P4 天真无邪
Innocence

P8 活力
Vitality

P9 欢悦
Delight

毛麻仓库 3F
Maoma Warehouse 3F

毛麻仓库 4F
Maoma Warehouse 4F

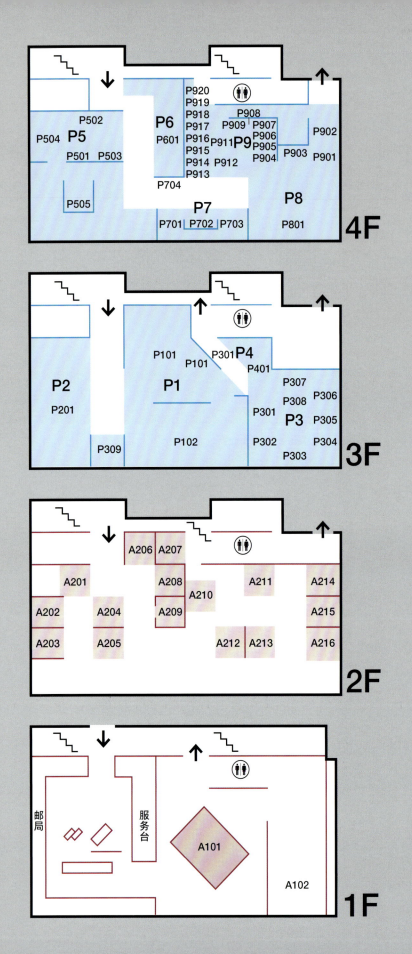

4F

P502
P504 P5
P501 P503
P505

P6
P601

P920
P919
P918
P917
P916
P915
P914
P913

P908
P909 P907
P906
P911 P9 P905
P912 P904

P902
P903 P901

P704

P7
P701 P702 P703

P8
P801

3F

P2
P201
P309

P101
P101
P1
P102

P301 P4
P401
P307
P308 P306
P301 P3 P305
P302 P304
P303

2F

A206 A207
A201
A208
A210
A202 A204 A209
A203 A205

A211
A212 A213

A214
A215
A216

1F

邮局
服务台

A101
A102

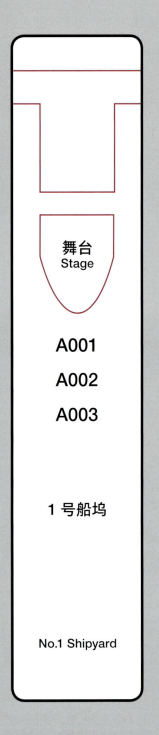

舞台
Stage

A001

A002

A003

1号船坞

No.1 Shipyard

Exhibition Infomation　展览信息

公共艺术作品
Public Artworks

由北川富朗先生邀请包括刘建华、沈烈毅、徐震、向阳、韩家英、浅井裕介、川添善行、高桥启祐、菲利斯·瓦里尼、理查德·威尔逊、埃斯特·斯托克、大岩·奥斯卡尔等数十位国际知名艺术家，结合杨浦滨江的历史底蕴、空间特点和未来展望，选择合适的场所或建构筑物，在地创作了约 19 件大型公共艺术作品，留存在 5.5 公里滨江空间中，成为见证当代艺术与市民日常生活融合共生的城市文化景观。

与此同时，本届空间艺术季还提供 5 个公共艺术品点位，向世界征集优秀作品。征集遴选出的优秀作品之一《点燃烟灰的生态可循环装置》也制作落成，留存于滨江空间内。

Fram Kitagawa invites dozens of internationally renowned artists, including Liu Jianhua, Shen Lieyi, Xu Zhen, Xiang Yang, Han Jiaying, Yusuke Asai, Yoshiyuki Kawazoe, Keisuke Takahashi, Felice Varini, Richard Wilson, Esther Stocker, and Oscar Oiwa, to create about 19 large-scale public artworks on site along the 5.5 kilometers of riverfront space. These artworks are created in suitable places or with structures in combination with the history, spatial characteristics, and future outlook of the Yangpu Riverfront, which have now become an urban cultural landscape for the integration of contemporary art and people's daily life.

Meanwhile, this year's SUSAS also provides five public artwork spots to collect outstanding works from the world. One of the outstanding works *Carbonized Earth Garden* selected from the call has also been completed and preserved in the riverfront space.

主展馆片区
Main Exhibition Area

上海自来水科技馆
Shanghai Water Technology Museum

黄 浦 江

TIANZHANG RD 天章路

TONGBEI RD 通北路

XUCHANG RD 许昌路

HUAIDE RD 怀德路

JIANGPU RD 江浦路

QIQIHAER RD 齐齐哈尔路

YANGSHUPU RIVER 杨树浦港

MEIZHOU RD 眉州路

ANPU RD 安浦路

DANDONG RD 丹东路

DESHA RD 德沙路

KUAN DIAN RD 宽甸路

YANGSHUPU RD

YUANSHEN RD 源深路

TAOLIN RD 桃林路

大 江 道 邑 昌 路 滨

MINSHENG RD 民生路

BINJIANG AVENUE

CHANGYI ROAD

2019 上海城市空间艺术季艺术作品分布与公共空间
Artworks' Locations in 2019 SUSAS

01
投射之窗 Projective Window
袁烽 Philip F. Yuan

02
天外之物 Extraterrestrial Object
刘建华 Liu Jianhua

03
方块宇宙 Square Universe
艾斯特·斯托克 Esther Stocker

04
城市的野生 Wildness Growing
Up in the City
浅井裕介 Yusuke Asai

05
点燃烟灰的生态可循环装置
Carbonized Earth Garden
李科文 / 百安木设计咨询（北京）
有限公司上海分公司
Li Kewen / Ballistic Architecture
Machine Desian Consultancy
(Beijing) Co., Ltd Shanghai
Branch

06
树 STREEG
帕斯卡尔·马尔蒂那·塔尤
Pascale Marthine Tayou

07
时间之载 Time Shipper
大岩·奥斯卡尔 Oscar Oiwa

08
山——索福克勒斯、赫拉克勒斯、
苏格拉底、荷马
Mountains — Sophocles,
Hercules, Socrates, Homer
徐震® XU ZHEN®

09
若冲园 Ruo Chong Garden
宋冬 Song Dong

10
拱门：艺术、旅行与相遇 Gates:
Art, Travel and Encounter
荷塞·德·吉马良斯 José de
Guimarães

11
一年 / 一万年 1 Year/10
Thousand Year
川添善行 Yoshiyuki Kawazoe

SONGPAN RD 松潘路

LINQING RD 临青路

NINGWU RD 宁武路

LONGCHANG RD 隆昌路

GUIYANG RD 贵阳路

NEIJIANG RD 内江路

DINGHAI RD 定海路

DINGHAI RD BRIDGE 定海路桥

黎平路

GUANGDE RD 广德路

杨树浦

PINGDING RD 平定路

TENGYUE RD 腾越路

FUXING ISLAND CANAL 复兴岛运河

BRIDGE

杨浦大桥

H U A N G P U   R I V E R

08
09
10
11
12
13
14
15
16
17

浦东大道

公共艺术作品
地图
Map of Public
Artworks

● 浦东滨江贯通点位分布
Pudong Riverside Connections

01

投射之窗
Projective Window

袁烽
Philip F. Yuan

创作年份：2019—2020
尺寸：5米 ×5米 ×10.76米
材质及结构：混凝土基础、钢
框架、铝合金杆件等
地点：芒草园

Year of Creation: 2019-2020
Size: 5m×5m×10.76m
Material and Structure:
Concrete foundation, steel
frame, aluminum alloy
components, etc.
Location: Miscanthus Garden

我们该如何"观看"当代大都市以及其中的地标建筑？是无意间的漠视，还是细微处的观察？《投射之窗》装置为我们提供了一个重新"看"城市的机会。无论是透过正交的网格对城市地标的凝视；抑或穿过彩色的装置场域，在漫步之间余光扫过城市天际，都在观者与城市之间的互动提供着契机，并不断地刺激着我们重新思考城市景观的日常意义。

How do we "see" the contemporary metropolis and its landmarks? Is it an indifferent visual-scanning or a subtle observation? *Projective Window* installation offers us an opportunity to "re-see" our city. Whether it's gazing at urban landmarks through an orthogonal grid-frame, or unconsciously perceiving across the city's skyline through a coloured field, the installation could continuously cultivate the dynamic interactions between the viewer and the cityscape, and it would constantly stimulating us to rethink the everyday meaning of the urban landscape.

袁烽
同济大学建筑与城市规划学院教授，博士生导师；曾担任麻省理工学院（MIT）客座教授、美国弗吉尼亚大学（UVA）"托马斯·杰佛逊"教席教授；上海创盟国际建筑设计有限公司创始人。专注于建筑后人文建构理论研究、数字化技术研发以及建筑设计实践，一直致力于推广数字化设计和建造理念在建筑学中的应用。已出版中英文著作 10 余本，多次受邀在国内外知名大学与国际会议讲座，设计展获国际、国家级各类奖项。袁烽也曾多次参加威尼斯建筑双年展、芝加哥建筑双年展等多个国际展览，多个作品被纽约现代艺术博物馆（MoMA）、香港 M+ 视觉文化博物馆永久收藏。

Philip F. Yuan
A professor in College of Architecture and Urban Planning (CAUP) at Tongji University, visiting professor at Massachusetts Institute of Technology (MIT), Thomas Jefferson professor at University of Virginia (UVA), the founder of Archi Union Architects. His research mainly focuses on the fields of Post-human tectonics, robotic fabrication and architectural design practices, promoting the application of digital design methodologies and fabrication theories in the discipline of architecture. Philip F. Yuan has published more than 10 books on his work in both English and Chinese. He has been invited to give keynote speeches and lectures at various institutes both domestic and overseas, His research and projects have received many international awards. He attended multiple international exhibitions including the Venice Biennale of Architecture and the Chicago Architectural Biennial. His works have entered several renowned museums including the permanent collections of the Museum of modern Art (MoMA), New York and the M+Museum of Visual Culture, Hong Kong.

02

天外之物
Extraterrestrial Object

刘建华
Liu Jianhua

类别：雕塑类艺术品
创作年份：2015—2019
尺寸：20 米
材质：不锈钢、夜光漆、镭射灯
地点：墙头草场

Type: Sculpture
Year of Creation: 2015-2019
Size: 20 m
Material: Stainless steel,
luminous paint, laser light
Location: Defene Lawn

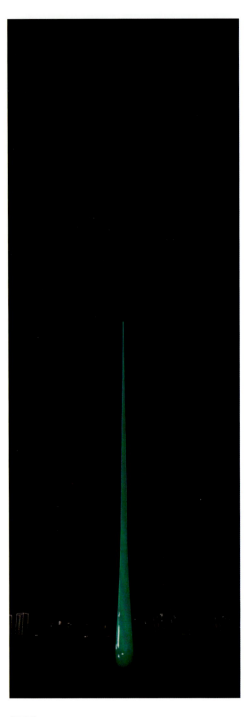

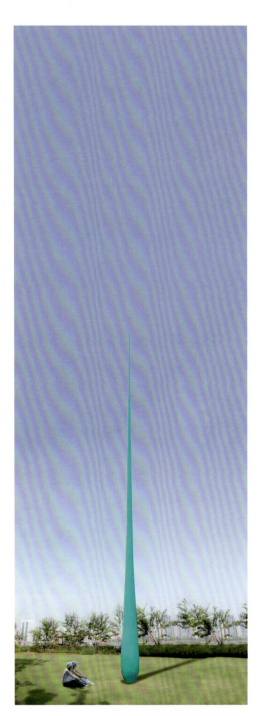

效果图
design sketch

刘建华

1962 年出生，现生活和工作在上海，他以综合材料为主要媒介，是中国当代艺术领域最具实验性、代表性的艺术家之一。1989 年开始尝试在当代背景下进行实验性的创作。刘建华的作品是对近年来不断困扰中国的经济和社会变化所做出的反映。2008 年，他一反先前多年对全球化及中国社会急剧转型引发的诸多问题、焦点的近距离直接关注，提出"无意义、无内容"的概念来进行创作，从 2008 年作品《无题》开始了一个全新方向的探索，并形成了当代艺术创作的个人语言体系。

Liu Jianhua

Born in 1962. Now lives and works in Shanghai, China. He is one of China's best-known contemporary artists who experiment with comprehensive materials. In 1989, within a contemporary context, he started his own experimental practices. His porcelain and mixed media works reflect the economic and social changes in China as well as the problems that follow suit. In 2008, he shifted his previous close and direct attention on the problems emerged in China from globalization and sharp social changes to "no meaning, no content", which declared the fairly new exploration of his creation with works *Untitled* in 2008 and therefore has developed his own expression system for contemporary art.

该项目的设计场地位于防汛墙后部的三块具有较高视野的绿化草坪上，长度分别为 26.5 米、64 米和 24 米，最大宽度约 15 米。

该区域曾为上海第一毛条厂，目前已改造为滨水公共空间，供人们休憩、停留。该区域的景观设计运用海绵城市理念，建设了雨水湿地生态示范点。

作品《天外之物》选择了从天而降的雨滴造型，造型抽象、安静、纯粹，从自然的形态中抽离出来，同时与艺术家的重要作品《迹象》及个人艺术语言相关。单纯抽象的雨滴形态，会让人感受到自然与生活及艺术的密切关联，人们在当今快速工作生活之余，可体会艺术带来的轻松、愉悦的感觉。

作品的颜色采用荧光材料，通过光能的作用，使人们在晚上也能看到该艺术作品的形态，在造型顶端置放一个激光灯，晚上激光灯直射天空，长长的激光射线让人感受到水滴造型似从天而降。

The design site is located on three green lawns with high field of view behind the flood control wall, with lengths of 26.5 m, 64 m and 24 m, respectively, and a maximum width of about 15 metres.

This area used to be the first wool top factory in Shanghai, which has been transformed into a waterfront public space for people to rest and stay. The landscape design of the area uses the concept of sponge city to build a rainwater wetland ecological demonstration point.

The work employs the shape of descending raindrops, constructing an abstract, quiet and pure design, which is from the natural shape of a raindrop. Meanwhile, this work is related to the artist's renowned work *Trace* alongside his personal artistic language. The pure and abstract shape of raindrops will draw people closer to the interconnection between nature, life and art, creating a relaxed and pleasant feeling to those who live under a fast, pressured life.

This piece of work adopted colours made of fluorescent materials. Through luminous energy emitted by the material, the work can be seen clearly even at night. A laser light is placed at the top of the artwork, radiating directly into the night sky. The long laser ray will make the raindrops seem like they are falling from the sky.

03

方块宇宙
Square Universe

埃斯特·斯托克
Esther Stocker

类别：装置艺术品
创作年份：2019
尺寸：可变尺寸
材质：复合铝板
地点：渔货廊架

Type: Installation
Year of Creation: 2019
Size: Variable size
Material: Composite
aluminum plate
Location: Fishing Corridor

该区域曾为上海第一水产批发市场的渔货廊架，约110米长、12米宽，地面为木制架空平台。

该水产批发市场可追溯至1945年建立的中国海洋渔货市场，目前已改造成公共空间以及配套服务使用。廊架空间的南侧保留原有防汛墙以及6道闸门，通过闸门可与外侧码头相连接。场地地面荷载不大于0.5吨/平方米，艺术品的设置应结合以上因素考虑，同时需兼顾建筑本身的经营性需求。

正方形是一种几何基本形式，也是一种世界符号。在古代中国，地球或整个宇宙被认为是正方形的。正方形是一种强大而又平静的形状，因为它边长相等，它本身就是一种逻辑形式——它和人世间的存在、力量和平静联系在一起，也可以说它象征着现实，立柱在平面图上也是方形。最重要的是，这种几何式可以在自由浮动的组成形式中同时带来平静和神秘。它并非封闭的形式，而是一种开放的过程和进步的能力。

不同大小的黑色方块的组成方式创造了失序与秩序，而拱廊现有构造与浮动正方形间的张力和关系，代表了已知与未知之间的转换和相遇，也是现实中的神秘标志。

此作品让观赏者在穿越拱廊结构的时候，就像进入了一个几何宇宙。观赏者潜入艺术作品并参与其中，体验美感和空间关系。在装置艺术中移动、改变视角的同时，也让人在心中对现实、生活有了不同的看法。

This area used to be the fishing corridor of The First Aquatic product wholesale Market in Shanghai, with a width of about 12 m and a length of 110 m. The ground is wooden raised platform.

The aquatic products wholesale market can be traced back to the Chinese Marine fishery Market established in 1945, which has been transformed into a public space and supporting services. On the south side of the gallery space, the original flood control wall and 6 gates are maintained, through which they can be connected with the outer wharf. The ground load should not be greater than 0.5 t/ ㎡. It is suggested that the setting of artworks should be considered in combination with the above factors, while taking into account the operational requirements of the building itself.

Prevailing to all, is the theme of the square. The square is a geometric basic form but also a world symbol. In ancient China, the earth, the whole cosmos was considered quadratic. The square is a powerful but also calm form, because its lenghts are all equal, it is a logical form that is resting in itself. It associates with earthly existence, power and calmness. One could say it symbolizes reality. It also refers to the ground plan of the existing columns which are in square form. Essential to the artistic use of the square is the fact that this geometric form is able to bring calmness and mystery at the same time in a free-floating organization of forms. The work is less understood as a closed form, but rather an open process and the ability of progress.

Black squares in different dimensions are composed to create complexity and order. The tension and relation between the existing structure of the arcade and the floating squares represent shifts and encounter between the known and the unknown. It could be read as a sign for accepting mystery inside reality.

The principal idea originating in the work is the experience of the viewer to enter a square geo- metric universe when crossing the structure of the arcade. Visitors submerge and participate in the art- Work experiencing the aesthetic sensibility and spatial relation. Moving through the installation requires a change of perspective. Psychologically, it means gaining a different perspective on reality, on life.

埃斯特·斯托克

1974 年生于意大利的西兰德罗，之后在维也纳、米兰、帕萨迪纳学习艺术，现居奥地利维也纳。其个展包括：2019 年，布达佩斯，Varfok 画廊《进入混乱——复杂的几何图形》；2014 年，乌尔姆艺术协会《疯狂的几何学》；2013 年，捷克鲁德尼斯现代艺术画廊《无限空间》；2012 年，李特尔博物馆《失序的肖像》；2011 年，巴黎 Alberta Pane 画廊《肮脏的几何》；2011 年，罗马当代艺术博物馆《共同的命运》；2010 年，Ko. Ji. Ku 协助，热那亚 Studio 44 画廊《歌剧的孤独（布兰乔）》；2010 年，维也纳 Krobath 画廊；2008 年，伦敦 52 博物馆《我所不知的空间》、维也纳路德维希现代艺术基金会《幸福的几何》。

Esther Stocker

Born 1974 in Silandro, Italy. Art studies in Vienna, Milano, Pasadena. Living and working in Vienna, Austria. Her solo shows include—2019: *Entering the Mess — Perplexing Geometries* (with Tamás Jovanovics), Várfok Gallery, Budapest; *Anarchy of Forms*, drj art projects, Berlin; 2014: *Verrückte Geometrie*, Kunstverein Ulm; 2013: *Unlimited Space*, The Gallery of Modern Art in Roudnice nad Labem, Czech Republic; 2012: *Portrait of Disorder*, Museum Ritter, Waldenbuch; 2011: *Dirty Geometry*, Galerie Alberta Pane, Paris:*Destino Comune*, Macro, Roma; 2010: *La solitudine dell' opera (Blanchot)*, Associazione Ko.Ji.Ku., Galleria Studio 44, Genova Galerie Krobath, Wien; 2008: *What I don' t know about space*, Museum 52, London; *Geometrisch Betrachtet*, Museum Moderner Kunst Stiftung Ludwig, Wien.

埃斯特·斯托克：渔货廊架是一个很美的场地，但也相当复杂。它有许多空间，很难找到一个中心点。在拱廊内部有一点私密性，人既被包围，又与城市相连通。

我们打造了一个整体，不只是单个图形，也是一个结构，类似于肌理。它仿佛是一个由方块组成的宇宙，就像与城市肌理平行的图像。我觉得生活也是这样，城市或建筑是结构，我们穿梭其中，但同时头脑中又有一个与之平行的肌理，就像思维一样。

当你看作品局部的时候，仿佛那是一个只会存在于脑海里的图像。因此，它是一个既好玩又非常有体验感的空间。

过去与当下的连接是我感兴趣的一点。历史中包含着当代的征兆。艺术家与建筑之间发生交流，现有的形态与改造的样貌之间也有沟通。我坚信图形大小的变化能够起到沟通的作用。我们不一定要丢掉建筑原本的个性，只需要增加新的体验来阐释。

Esther Stocker: "Fisherman's Corridor" is a beautiful place, but also quite complex. It has a lot of space and hard to find a central point. There is a bit of privacy inside the arcade, where people are surrounded and connected to the city.

We created an entirety, not just a single figure, but a structure, like a kind of texture. It is as if a universe consisted of squares, like an image parallel to the city fabric. I think it's the same as life. A city or a building is a structure, when we move through it, at the same time there is a parallel texture in our mind, just like a thought in mind.

When you look at parts of the artwork, it looks like an image that only exists in your mind. As a result, it is a playful space with full experiences.

I'm interested in the connection between the past and the present. History contains the omens of the present. There are communications between the artist and the building, between the existing form and the appearance after renovation. I believe that changing the size of graphics will create communications. We don't have to lose the original characteristics of the building, just add new experiences to illustrate it.

04

城市的野生
Wildness Growing Up in the City

浅井裕介 Yusuke Asai

类别：绘画类艺术品
创作年份：2019
尺寸：一组两件，每件 95 米
×20 米
材质：溶着性白线橡胶
地点：打捞局码头

Type: Painting
Year: 2019
Size: 95mx20m x2
Materials: Soluble white
rubber
Location: Salvage Bureau Pier

该场地位于打捞局码头，约 200 米长、20 米宽，场地地面为混凝土铺装，是滨水空间中最为开阔的一段。

该码头区域一直作为救助打捞作业码头，目前已改造为公共空间，但依然兼有救捞作业的功能。在策划该场地艺术品时，考虑到该码头仍具有潜在的打捞作业需求，故采用了大地彩绘这种无地面突出物的艺术形式。

人、自然、物、历史、过去现在未来，各种各样的东西在这里相遇，存在着无比的生命力，而艺术家希望将这种生命力转换为可视化的作品。这件作品不是将新制作的物品搬来此地，而是将刺青般的画刻在场所中。

作品使用象征人工制品的斑马线白色橡胶材料，制作植物象征的花朵，移动生物象征的鸟类，透过它们来探讨在城市中自然的存在方式。它刺激了人们的想象力，重新思考此场域，也是一个无论何时都能如公园般让大众享受的作品。艺术家的制作手法比起独立自行，更多采用了与市民及助手们一同工作的方式，来开拓这土地上的可能性，注入更多的生命力，孕育新的绘画。

The site is located at the Salvage Bureau Wharf, about 200 m long and 20 m wide. The ground is paved with concrete, and here is the most open section of the waterfront space.

This dock area has always been used as a salvage operation dock. Now it has been transformed into a public space, but it still has the function of salvage operation. During the planning of artworks at this site, considering that the wharf still has potential salvage operations requirements, the art form of land painting without ground protrusions was adopted.

People, nature, objects, history, the present and future, all kinds of things meet here, producing an incomparable vitality, and the artist wants to transform this vitality into a visual work. This work is not to move the newly made items here, but to engrave the tattoo-like paintings in the place.

The work uses zebra-line white rubber materials that symbolize artifacts, making plant-symbolized flowers, and moving creature-symbolized birds to explore the natural ways of existence in cities. It stimulates people's imagination, encouraging them to rethink this field. It is a work that can be enjoyed by the public all the time like a park. Compared with the independent creation method, the artist prefers to work with the citizens and assistants to develop the possibilities of the land and inject new vitality and new paintings.

浅井裕介
1981 年出生于日本东京，1999 年神奈川县立上矢部高等学校美术陶艺科毕业。2003 年开始创作 "Masking Plant" 系列，使用耐水性马克笔在胶带上绘制植物。"泥绘" 系列利用创作驻留地的土壤和水绘制动物和植物。"成为白线的植物" 系列则将柏油路的白色标线切割出动植物的形态后以火进行烧制。其作品饱含生命力，可以存在于各类环境和场景。近年来多次发表超过 10 米的大型 "泥绘" 作品，备受瞩目。浅井将动植物铺满于画布，大型动物的体内时而嵌套着小小的动物，描绘着微观中存在宏观的宇宙生态。

Yusuke Asai
Born in 1981 in Tokyo, Japan. In 1999, he graduated from Ceramic art course, Kamiyabe High School, Kanagawa, Japan. He has been making Masking Plant series since 2003, which are permanent marker drawings on masking tapes that are freely pasted on the walls and floors. He collects soil and water from the local area and creates Earth paintings series. He also creates Sprouted Plants series that he cuts out the shapes of animals and plants from the sheet of road marking white line and torches on the road surface. His works are full of vitality and can exist in various environments and scenes. Recently, he has received a lot of attention for large-scale Earth Painting seires that are over 10m, He juxtaposes animals and plants without any space, small animals and plants appear inside of bigger living things just like macro exists inside of micro, the ecosystem in the universe.

浅井裕介：不只是对艺术感兴趣的人，还有很多每天经过这条路的人，都参与到剪切作品的过程中。其实把画笔交给他人这件事，有时候很让我害怕，但也出现了令人惊喜的东西。我看到小孩想要像大人一样画画，大人像小孩一样画画，非常新鲜有趣。

正是这样的相遇，在不停地改变着作品的样貌。我每天都会修改计划表，这取决于来访的人、天气，以及与场所的交流，感觉像是在这里孕育一个生命。

在这个世界上，想要快乐幸福地生存，就要去考虑与自然之间的关系。因此在这个空间里，我想创造一个像森林那样的不会令人厌烦的东西。在画上来回走时，有时脑海中把所见到的元素组合在一起，会感到期待又兴奋，这样很健康，正是我所说的 "生命力" 和 "野生"。这种欣欣向荣的感觉真好。

人类最初创作的那些壁画，是能与所在场所和环境很好地契合的。就像食用从田里采摘的蔬菜时，因为夹杂着 "移动" 这样的行为，就很不自然，也会与实际产生偏差。所以对于不能 "移动" 的东西，假如不去到那个地方，就看不到它，这是土地所蕴含的力量所在。我相信这一点，也是这样去进行艺术实践的。

Yusuke Asai: Not only people who are interested in art, but also many people who pass through the road every day, are involved in the process of cutting the work. In fact, it was scary to give way my brushes to others, but there was something amazing about it. I see kids trying to draw like adults, and adults drawing like children, which is very new and interesting.

It is this kind of encounter that constantly changing the appearance of the artwork. I revise the schedule every day, depending on the visitors, the weather, the interaction with the site, and it feels like I am incubating a life here.

If we want to live happily in this world, we have to consider our relationship with the nature. In this space, I wanted to create something like a forest that could not be boring. As you walk back and forth on the painting, you feel anticipation and excitement when you combine every element you see in your mind. This is very healthy, and what I call "vitality" and "wildness". It feels good to see this thriving scene.

The first murals created by human beings fit in well with the place and environment. For example, when eating vegetables picked from the field, it is not natural because of the behavior like "transport", and also deviated from reality. If something can't be "transported", then you won't be able to see it unless go to that place. This is the power of the land. I believe in this and carry out my artistic practice in this way.

05

点燃烟灰的生态可循环装置
Carbonized Earth Garden

李科文／百安木设计咨询（北京）有限公司上海分公司
Li Kewen / Ballistic Architecture Machine Desian Consultancy (Beijing) Co., Ltd. Shanghai Branch

本作品为征集遴选优秀作品
This is the work selected from the call

创作年份：2019
材质：不锈钢、油漆
尺寸：6.4米（长）×1.5米（宽）
×3.8米（高）
地点：绿之丘

Year of Creation: 2019
Material: Stainless steel,
Paint Size: 6.4(L)m×1.5(W)m
×3.8(H)m
Location: The Green Hill

场地原为烟草公司仓库用地，现已被改造为生态型屋顶花园，雕塑利用有趣的烟头造型为场所的历史与生态花园的现状带来相遇的可能。雕塑灵感源自城市街道上被随意丢弃的烟头，而"烟草"在正确利用的前提下其成分却是具有生态价值的，折断烟头雕塑的设计意图正是在探索这种"烟"的双重现实。

This site was originally used as a warehouse for a tobacco company but now it has been renovated into an ecological roof garden. Our sculpture uses interesting cigarette butts to bring the possibility of meeting between the history of the site and the existing ecological roof garden. The form is inspired by cigarette butts that are randomly discarded on city streets, while the ingredients of tobacco have ecological value under the premise of creative reuse. The design intention of the broken cigarette sculpture is to explore the meaning of "cigarette" in historical and ecological realization.

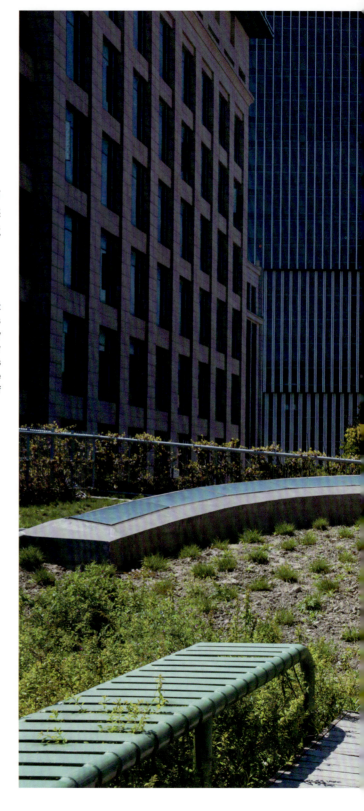

06

树
STREEG

帕斯卡尔·马尔蒂娜·塔尤
Pascale Marthine Tayou

创作年份：2019—2020
尺寸：高 6.5 米，树冠直径约
6.5 米
材质：铜、石
地点：水杉花园

Year of Creation: 2019-2020
Size: The height is 6.5 m, and
the diameter of the canopy is
about 6.5 m
Material: Bronze, stone
Location: Metasequoia
Garden

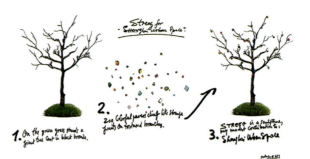

变革是对梦想的征服。它们拥有世界性的特质而非仅仅占有任何特定的领域。作品《树》是代表着我们所有为争取自由而战的一颗树，是艺术家对人类及人类所创造的所有变革的歌颂。

首先用青铜在草地上铸造一颗大树。各种颜色的 200 个铺路石块像奇形怪状的水果一样黏在扭曲的树杈上，它们共同组合形成这一雕塑。

Revolutions are dream conquests. They have no territory other than universal geography. STREEG is the tree of all our battles for freedom. It is the artist's tribute to mankind and all its revolutions.

Firstly, to cast a big tree in the grass with bronze. Then, 200 coloured paving-stones are stuck to the twisted branches like grotesque fruit, and they combine to form the sculpture.

帕斯卡尔·马尔蒂那·塔尤

1966 年生于恩康桑巴，在比利时根特和喀麦隆雅温得生活和工作。自 20 世纪 90 年代初以来，艺术家参加 2002 年第 11 届卡塞尔文献展，以及 2005 年、2009 年威尼斯双年展后，帕斯卡尔·马尔蒂那·塔尤已为国际群众所熟知。其艺术创作不局限于某种媒介或主题，具有可变性。尽管主题可能千差万别，但它们都以艺术家本人为出发点。在艺术家的职业生涯的开端，他就在名字里加上了字母"e"，以赋予其一个女性化的结尾，从而讽刺地使自己脱离了艺术原创的重要属性及性别归类。

Pascale Marthine Tayou

Born in Nkongsamba in 1966. Lives and works in Ghent, Belglum and in Yaoundé, Cameroon. Ever since the beginning of the 1990's and his participation in Documenta 11(2002) in Kassel and at the Venice Biennale (2005 and 2009), Pascale Marthine Tayou has been known to a broad international public. His work is characterized by its variability, since he confines himself in his artistic work neither to one medium nor to a particular set of issues. While his themes may be various, they all use the artist himself as a person as their point of departure. Already at the very outset of his career, Pascale Marthine Toyou added an "e" to his first and middle name to give them a feminine ending, thus distancing himself itonically from the importance of artistic authorship and male/female ascriptions.

07

时间之载
Time Shipper

大岩・奥斯卡尔
Oscar Oiwa

类别：装置艺术品
创作年份：2019
尺寸：15.8米×3.5米×2.5米
材质：玻璃、混凝土、钢、泥土、树木
地点：技能草场

Type: Installation
Year of Creation: 2019
Size: 15.8m×3.5m×2.5m
Material: Glass, concrete, steel, soil, tree
Location: Skills Lawn

场地选址区域位于永安栈房西侧约 80 米长、60 米宽的斜坡草地。永安栈房为文保建筑，建于 20 世纪上半叶，曾经是永安百货公司的滨江仓库，后又划归上海化工厂使用。今后，西楼将改造成为世界技能博物馆。

在上海的历史中，多种多样的文化、技术、人群自大海、河川、港口流入这片土地。为了让人们联想到时间和历史，艺术家希望为杨浦滨江的艺术项目创作一艘能够运送"时间"的隐形巨轮。制作一个透明的容器，放入上海的土壤，呈现出地层的结构。经过数年，这里会长出花草，或许有一天还会长出树木。很难想象百年之后的上海会是什么样子，但艺术家创作的初衷就是把当下的作品与今后的人们建立联结，向未来起航。

The site is located at the grassland slope which is about 60 m wide and 80 m long on the west side of Wing On Warehouse. Wing On Warehouse is a cultural security building, built in the first half of the 20th Century. It used to be the riverside warehouse of Wing On Department Store, and then it was put into use by Shanghai chemical plant. In the future, the west building will be transformed into a World Skills Museum.

In the history of Shanghai, a variety of culture, technology, people from the sea, rivers, ports into this land. In order to remind people of time and history, the artist hopes to create an invisible ship that can transport "time" for the art project of Yangpu Waterfront. Make a transparent container and put it into the soil of Shanghai, showing the structure of the stratum. After several years, flowers and plants will grow here, and maybe trees will grow one day. It's hard to imagine what Shanghai will be like in a hundred years, but my original intention of creation is to connect the current works with the future people and set sail for the future.

大岩·奥斯卡尔

1965 年生于巴西圣保罗，1989 年毕业于圣保罗大学建筑学系。从建筑学系毕业后，他在东京的建筑事务所工作的同时，也作为艺术家活动。于 2002 年移居纽约，现以美国为据点活动。奥斯卡尔的作品将其充满故事感的世界观在油画画布上强有力地展现。凭借其独特的幽默感和想象力，奥斯卡尔旅居圣保罗、东京、纽约多地并持续进行创作。他经常旅行，在移动的过程中探索其扎根于多种文化的自我认同。他使用细致的描绘和鸟瞰式的构图，在新闻报道和网络中发掘社会问题，在仔细调查后将其展现在巨大的画面上，其作品被国内外众多美术馆收藏。

Oscar Oiwa

Born in Sao Paulo, Brazil in 1965, graduated from the Department of Architecture at the University of São Paulo in 1989. After graduating from the Department of Architecture, he worked as an artist at the same time as he worked in an architectural firm in Tokyo. In 2002, he moved to New York where he actually lived and worked. Oscar's work shows his worldly story full of stories on the canvas. With its unique sense of humor and imagination, Oscar traveled to and from São Paulo, Tokyo, and New York. He travels a lot and explores his self-identity rooted in multiculturalism as he moves. He uses detailed depictions and bird's-eye composition to discover social issues from news reports and networks. After careful investigation, he displays them on a huge screen. His works are collected by many museums around the world.

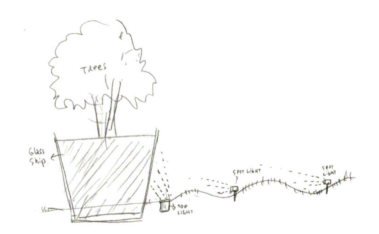

08

山——索福克勒斯、赫拉克勒斯、苏格拉底、荷马

Mountains — Sophocles, Hercules, Socrates, Homer

徐震 XU ZHEN®

类别：雕塑类艺术品
创作年份：2019
尺寸：3.26 米 ×2.45×
2.77 米，3.27 米 ×1.67×
1.69 米，2.43 米 ×1.94×
1.56 米，4.25 米 ×2.4×
2.72 米
材质：石膏、树脂
地点：值亭

Type: Sculpture
Year of Creation: 2019
SIZE: 3.26m×2.45m×2.77m,
3.27m×1.67m×1.69m,
2.43m×1.94m×1.56 m,
4.25m×2.4m×2.72 m
Material: Gypsum, resin
Location: Duty Pavillion

徐震®

艺术家、策划人、没顶公司创始人。1977 年出生，工作和生活于上海。徐震是中国当代艺术领域的标志人物，2004 年获得中国当代艺术奖（CCAA）"最佳艺术家"奖项，并作为年轻的中国艺术家参加了第 49 届威尼斯双年展（2001）主题展。徐震的创作领域非常广泛，包括装置、摄影、影像和行为等。他的作品在世界各地的博物馆和双年展均有展出，包括威尼斯双年展（2001，2005）、纽约现代艺术博物馆（2004）、国际摄影中心（2004）、日本森美术馆（2005）、纽约现代艺术博物馆 PS1（2006）、英国泰特利物浦美术馆（2007）、英国海沃德画廊（2012）、里昂双年展（2013）、纽约军械库展览（2014）、上海龙美术馆（2015）、卡塔尔 Al Riwaq 艺术中心（2016）、悉尼双年展（2016）、纽约古根海姆美术馆（2017）、沙迦双年展（2019）、洛杉矶当代艺术博物馆（2019）等。

XU ZHEN®

Artist, Curator, Madeln Company Founder. Born in 1977, he lives and works in Shanghai. Xu Zhen has been considered as an iconic figure in Chinese contemporary art. In 2004, Xu won the prize for Best Artist at the China Contemporary Art Award, and participated in the theme exhibition of the 49th Venice Biennale (2001). His practice covers various media such as installation, video, painting and performance, etc. Xu Zhen has exhibited internationally, at museums and blennales, such as, Venice Biennale (2001,2005), The Museum of Modern Art (New York,2004), Mori Art Museum (Tokyo,2005), MoMA PSI (New York, 2006), Tate Liverpool (2007), Hayward Gallery (Lonon, 2012), Lyon Biennial (2013), Armory Show (New York, 2014), Long Museum (Shanghai 2015). Al Riwaq Art Centre (Qatar, 2016), Sydney Biennial (2016), Guggenheim Museum (New York, 2017), Sharjah Biennial (2019), The Museum of Modern Art (Los Angeles, 2019), among others.

作品《山》取自古哲学家（索福克洛斯、赫拉克勒斯、苏格拉底、荷马）头像的局部，艺术家巧妙地将雕塑的胡子部分放大并倒放在空间内。作品的缺失部分引发了观众对它究竟是何物的想象，倒置的哲学家头像雕塑远看像中国的假山，仿佛是将自然景观移至室内空间。

艺术家在创造着新的经验的同时也挑战着观众的认知习惯。人们的认知在不断被城市影响的同时又对城市产生新的塑造。将《山》置于上海这样一座接纳了不同人群、融合了多元文化的大都市，将引发身处上海的观众对于自我认知及自身和城市关系的思考。

The artwork *Mountains* consists of elements sourced from portrait sculptures of ancient philosophers (Sophocles, Hercules, Socrates, Homer). The artist skillfully enlarged part of the beards of the sculptures and present them reversed in the space. The missing part of the works stimulates the endless imagination of the viewers on what they exactly could be. From a remote distance, the reversed philosophers' heads sculptures evoke Chinese stones, as if natural landscapes had been moved into an indoor space.

The artist creates new experiences while challenging viewers' cognitive habits. There is a reciprocal influence between the city and the people living in it, as they grow together. People's cognition is constantly influenced by the city, generating new aspects. With *Mountains* in Shanghai-a multicultural metropolis-viewers will be triggered to think about self-awareness and consider the relationship between the self and the city.

09

若冲园
Ruo Chong Garden

宋冬
Song Dong

创作年份：2019—2020
尺寸：12.6 米 ×6.6 米 ×6.5
米，6.6 米 ×4.2 米 ×5 米，
4 米 ×3 米 ×4 米
材质：钢、木、玻璃、日用品
地点：印记花园

Year of Creation: 2019 – 2020
Size: 12.6m×6.6m×6.5m,
6.6m×4.2m×5m, 4m×3m×4m
Material: Steel, wood, glass,
daily necessities
Location: Imprint Garden

该区域为长 42 米、宽 32 米的生态绿地。此处曾属于电站辅机厂东厂，是国家电力行业的重要单位，如今将改造为生态雨水收集及沙滩亲子花园。

作品名《若冲园》语出《道德经》，意为"大盈若冲，其用不穷"。艺术家在一处有水塘的庭园中放置三个盆景的基座，用钢框架和收集来的旧门窗构成一组看似盆景的景观系统。这些方体构成山形结构，其中放入并展示附近居民的日常物、工厂原地的旧物、废弃物。

盆景介乎于抽象与具象之间，看上去像是低像素的马赛克构成图，也像是城市建筑的天际线景观，又像是一船运输的货物。艺术品与现有场地的环境融为一体又相互映衬，共同构成一座当代园林，将上海这座国际大都市的历史文脉、当代发展与民众的情感和温度集于一处，使之相遇、对话、思考并面向未来。

The site is a 42 m long and 32 m wide ecological green area. The area used to belong to the east plant of the auxiliary plant of the power station and was an important unit of the national power industry. Now the area will be transformed into an ecological rainwater collection and parent-child beach garden.

The work's name, *Ruo Chong Garden* comes from *Tao and Teh*, which means "great surplus, its use is endless". The artist placed three bonsai pedestals in a garden with a pond, and used steel frames and collected old doors and windows to form a set of bonsai-like landscape systems. These cubes form a mountain-shaped structure, which puts and displays daily objects of nearby residents, old objects and wastes from the factory site.

Bonsai is between abstraction and figuration. It looks like a low-pixel mosaic composition, like the skyline view of city buildings, and also like a cargo transported by a boat. The artwork and the environment of the existing site are integrated and complement each other to form a contemporary garden, which combines the historical context of Shanghai as an international metropolis with contemporary development, the emotions and temperatures of the citizens, creating encounters and dialogues between them, facing and thinking about the future.

宋冬

1966 年生于北京，1989 年毕业于首都师范大学美术系，生活和工作在北京。现为中央美术学院、北京电影学院、广州美术学院客座教授。宋冬从早期的中国先锋艺术运动中脱颖而出，成为中国当代艺术发展中具有国际影响力的重要艺术家。艺术形式横跨行为、录像、装置、雕塑、摄影、观念绘画、戏剧和策划等多个领域，对人类行为短暂性的观念和艺术就是生活的理念进行了探索。用"无界"的态度进行创作和生活，使艺术融入生活中。先后参加了德国卡塞尔文献展、意大利威尼斯双年展、巴西圣保罗双年展、韩国光州双年展等众多国际艺术展。

Song Dong

Born in Beijing in 1966. Graduated from the Fine Arts Deportment of Capital Normal University in 1989. Lives and works in Beijing. Currently is a visiting professor at the Central Academy of Fine arts, Beijing Film Academy and Guangzhou Academy of fine arts. Song Dong stood out from the early Chinese avant-garde art movement and became an important artist with international influence in the development of Chinese contemporary art his art form spans multiple fields such as performance, video, installation, sculpture, photography, conceptual painting, drama and curating. He explores the concept of the transient nature of human behavior and the idea of art is life. With an unbounded attitude towards creation and life, he integrates art into his life. He has participated in many international art exhibitions such as Kassel Documenta (Germany), Venice Biennial (Italy), Sao Paulo Art Biennial (Brazil), and Gwangju Biennale (South Korea).

10

荷塞·德·吉马良斯
José de Guimarães

拱门：艺术、旅行与相遇
Gates: Art, Travel and Encounter

创作年份：2019
尺寸：共4组，每组2件，每
件长约5m，高3~4m，厚0.3m
材质：不锈钢
地点：滨江栈桥

Year of Creation: 2019
Size: 4 groups, 2
pieces in each group;
each piece is about
(L)5m×(H)3~4m×(W)0.3m
Material: Stainless steel
Location: Riverside Trestle

吉马良斯为黄浦江所创作的艺术品的灵感来源于一位根植于非洲安哥拉的作家。艺术家对非洲世界的了解不仅来源于艺术品和启蒙舞蹈等多种多样的仪式，还通过收藏手工艺品。而除了非洲地区的藏品外，他的藏品还包括亚洲和拉丁美洲的艺术品。通过这种方式，艺术家根据对多种地区及文化的了解创建了一张当代文化的旅游云图。

此次创作的艺术品位于黄浦江岸——穿越上海城区的黄浦江现在是一条运输来自世界各地货物的水路；在过去，它也是一条传递来自其他人群、文化的创新发明和异域知识的路线。"相遇"在旅行中产生，是知识的基础。这也是吉马良斯建造"拱门"的原因。也就是说，门或入口连接着出发与到达的两端，从而形成一次"相遇"。这就是非洲和世界其他地区的本土文化成为由艺术先锋传递的当代性中组成部分的过程及方式。在艺术家的作品中，它被诠释为多种形状的解构体，正如其向观众显现的那样。

在 20 世纪 80 年代，吉马良斯一反当时强调雕塑重量和体积的理念，创作出扁平得几乎没有体积的雕塑。正如这件作品所显示的那样，他始终坚持彩色的、幽默的特征，雕塑形体从各个角度看都细长、独特且明晰。因此，这些雕塑作品代表着普遍意义层面的旅程和交错，以及艺术、文化和文明的交汇——不同世界、不同文化和不同语言之间的"相遇"。

正如艺术家所言，由出发和返回所组成的旅行也是美学的理想领域。这就是为何黄浦江的长途步道及其门廊可以为其他文化提供一个入口，或者在更深的象征意义上，允许从一种文化过渡到另一种文化。

The works that the artist created for Huangpu River were produced by an author whose artistic genesis is rooted in Africa (Angola). The artist discovered the African universe not only through its artis- tic production, its broadly diverse rituals such as initiation dances, but also through the collections of its artefacts. These collections also include Asia and Latin America. In this way, the artist built a travel atlas of contemporary culture based on knowledge he gained about cultures in a wide array of regions.

The works of art are located along the Huangpu River, which crosses the city of Shanghai, a river that today is a route for conveying merchandise from around the world. In former times, it was also a waterway for vessels that brought innovations and unfamiliar knowledge from other peoples and cultures. The travels transmit the "encounter", which is the foundation for knowledge. That is why the artist built what he called "gates", in other words, doors or entry points serving as a route between the point of departure and the point of arrival and creating an "encounter", This is how the indigenous cultures of Africa and of other world regions became part of the knowledge of modernity, conveyed by artistic vanguards, which in this work is interpreted as a deconstruction of shapes, as is revealed to the observer.

In the 1980s, the sculptures of the artist were very flat, almost without volume, contradicting the concept of weight and volume of sculptures characteristic of that period. He has maintained that concept, as can be seen in the works built here, with their chromatic and ludic features, with shapes that are slender and uniquely visible from various perspectives. These sculptural works thus represent treks, crossing points and artistic, cultural and civilisational crossroads in a universalistic sense-"Encounters" of worlds, cultures and languages.

The voyage, as the artist said, consists of departing and returning, but it's also the ideal territory of aesthetics. That's why the long walkway of the Huangpu River, with its porticos, allows and provides an entry point for other cultures or, better, symbolically allows a transition from one culture to another.

荷塞·德·吉马良斯

1939 年出生于葡萄牙吉马良斯市。作为最杰出的葡萄牙当代艺术家之一，吉马良斯以其包含绘画、雕塑、公共艺术等多样的艺术形式活跃于国际。他对非西洋文明圈的艺术互动表现出浓厚兴趣，除了将其吸纳进自身作品的主题，还作为非洲、拉丁美洲、中国等地区的原始艺术的收藏家而闻名。他还操刀了日本越后妻有大地艺术节的标识设计。

José de Guimarães

Born in 1939 in Guimarães in Portugal. As one of the most prominent Portugese contemporary artists, Guimarães presents his artworks internationally in diverse forms ranging from painting, sculpture to public art. With profound interest in interaction of art in non-Western world, Guimarães not only takes such perspective into his own creations but also known as an eager collector of primitive art from Africa, Latin America and China. The signposts that stand across Echigo-Tsumari region (Japan) were made by Guimarães.

1 年／ 1 万年
1 Year / 10 Thousand Years

川添善行
Yoshiyuki Kawazoe

类别：装置艺术品
创作年份：2019
尺寸：大小不一的一组锥体
材质：食盐、肥皂、FRP
地点：皂厂咖啡馆

Type: Installation
Year of Creation: 2019
Size: A set of cones in
different sizes
Material: Salt, soap, FRP
Location: Soapery Cafe

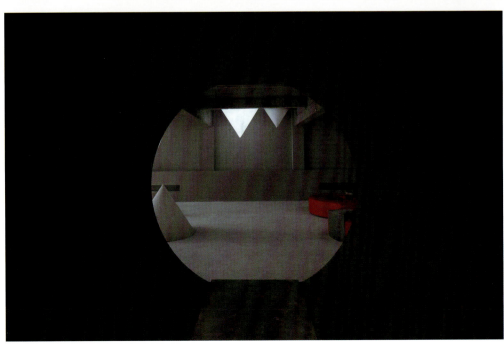

川添善行

建筑家，东京大学准教授。因设计长崎豪斯登堡的"奇怪的酒店"而载入吉尼斯世界纪录。担任大约时隔 100 年的东京大学新图书馆的设计，并于 2017 年建成。涉及领域不限于设计，还出版了《寻找隐藏在空间里的意识》（幻冬社）、《在这个城市生活》（彰国社）等著作。担任空间构想一级建筑师事务所，日本荷兰文化协会会长等要务。川添不只限于创作单体建筑，还进行了在和歌山县的加太渔村设立研究室分室等活动，对区域振兴做出的各种努力也引起了社会关注。

Yoshiyuki Kawazoe

Architect, Associate Professor at the University of Tokyo. His design of "Henn-na Hotel" in Nagasaki Huis Ten Bosch was recorded in the Guinness World Record. He designed the new library of the University of Tokyo 100 years after construction, and completed it in 2017. His fields involved are not limited to design, he also published *Looking for Consciousness Hidden in Space* (Gentosha Inc.), *Life in this City* (SHOKOKUSHA Publishing Co., Ltd.) and other books. He is the principal architect of kousou Inc. and the chairman of Japan and Netherlands Architecture Cultural Association. Kawazoe is not limited to the creation of building, but also has set up a research room sub-division in the Canadian fishing village of Wakayama Prefecture. His various efforts made in the regional revitalization have also attracted social attention.

该作品的场地位于曾经的老厂牌上海制皂厂所在地。上海制皂厂是上海唯一生产系列肥皂的国家一级企业，创建于 1923 年，号称远东最大制皂厂，如今将在制皂厂内生产水池的旧址上建造咖啡厅。同时，上海水乡也曾是生产优良食盐的产地。于是，艺术家利用"肥皂"和"食盐"两种纯白的物质，设计出像钟乳洞一样的空间，意图在空间中加入能让人感受到场地历史的装置。

艺术家以石笋为空间主题，利用肥皂与盐的化学反应，试图在此地呈现出钟乳石的成长过程。这个作品不仅是视觉上的，肥皂的气味还会刺激嗅觉，让观众体验到与历史相遇的感受。

厂房里遗留的半圆形水泥管道给艺术家留下了非常深刻的印象，从每一个管道中走过，都像是穿越了时空，这种时光的叠加又与艺术家所创作的作品《1 年 / 1 万年》理念不谋而合。随着上海梅雨季节的到来，水泥建筑上偶尔滴下的水滴，落在这些钟乳石上，空间中淡淡的肥皂香气，视觉与嗅觉的多重感受或许会让人不禁去思考时间这一永恒的哲学问题，也会在不经意间回想起这个空间的过去，联想它的未来。

The site of the work is located at the old factory brand Shanghai Soap Factory. He is the principal architect of kousou Inc. and the chairman of Japan and Netherlands Architecture Cultural Association. Shanghai Water Town was where fine salt was produced. The artist uses the pure white matter of "soap" and "salt" to design a space like a stalactite cave. The intention is to add a device that allows people to feel the history of the site in the space.

The artist takes stalagmites as space theme, using the chemical reaction of soap and salt, trying to present the growth process of stalactites here. This work is not only visual, the smell of soap also stimulates the olfactory experience, allowing the audience to meet the history.

The semi-circular cement pipes left in the factory gave a deep impression on the artist. Walking through each pipe seems to have traveled through time and space. This superposition of time coincides with the work's idea of *1 Year/10 Thousand Year*. With the arrival of Shanghai's rainy season, the occasional drops of water dripping from the cement buildings fall on these stalactites, the faint soap aroma in the space and the multiple sensations of sight as well as smell may make people cannot help thinking about the eternal philosophical question of time,and also inadvertently recall the past of this space and think of its future.

12

轻舟过隙
Boat of the Floating World

向阳
Xiang Yang

类别：雕塑类艺术品
创作年份：2019
尺寸：18.5 米（长）×0.8 米
（宽）×3.73 米（高）
材料：铸铜
地点：煤气厂公园西

Type: Sculpture
Year of Creation: 2019
Size: 18.5 (L)m ×0.8 (W) m×
3.73 (H) m
Material: Cast copper
Location: Gas Factory Park
West

向阳

1967 年出生于贵州铜仁市。1991 年毕业于中央工艺美术学院（现清华大学美术学院）壁画系。1994 年于中国艺术研究院进修。1998 年旅居美国，现工作于纽约和北京。曾于纽约艺术与设计博物馆、费城佩德布莱德艺术中心、费城艺术联盟、澳大利亚国家博物馆、圣路易斯 LUMINARY 艺术中心，以及北京中国美术馆、今日美术馆等艺术机构展出其装置等艺术创作。2010 年受上海世博城市未来馆邀约创作大型作品《和谐塔》。2015 中英文化年，其艺术项目《非常建筑——东方遇见西方》于伦敦泰晤士河上巡游展出。2018、2019 年参加越后妻有大地艺术节和濑户内国际艺术节。其作品被英国达拉姆大学、中国美术馆、今日美术馆等机构永久收藏。

Xiang Yang

Born in Tongren, Guizhou, China in 1967, graduated from Central Academy of Design and Fine Arts of China in 1991, Xiang Yang continued his study in the Graduate School of China Art Academy in 1994. He moved to the United States in 1998.Now Xiang has lived and worked both in New York and Beijing. He has exhibited works at venues including Museum of Art and Design (NYC), Painted Bride Art Centre (Philadelphia), Art Alliance (Philadelphia), Today Art Museum (Beijing)and National Art Museum of China (Beijing). His large-scale work, *Harmony Tower*, was commissioned for Future Pavilion, EXPO 2010, Shanghai. In 2015, his art project *Ultra Architecture East Meets West* was exhibited along Thames River, London. In 2018 and 2019, XIANG participated Echigo-Tsumari Triennale and Setouchi Triennale. His work is in the permanent collection of art institutions including National Art Museum of China, Today Art Museum and Durham University.

舟，在东西方文化中各有着强烈的象征寓意：在佛家学说中，它渡世人远离苦海；在道家文化里，它是出世的载体；许多西方的宗教故事里也描述了挽救世界生灵的方舟。在艺术家的情感中，船蕴含着一种温暖如母亲怀抱的力量，载人去寻找精神和理想所向往的彼岸。艺术家旨在黄浦江畔创作一件"漂浮"在半空中的舟形建筑雕塑，与往来船只繁忙的黄浦江面形成一种动静、虚实空间的对比。

作品上半部分的"船屋"，呈现一座当代建筑结构的空间，希望观众能借此回味和想象过去曾在江岸边的船坞和老上海的建筑。观众随着行动视角的变化，可以看到这个建筑结构不同维度的视觉透视，由此，作品提供了不同的角度来看待建筑、文化和传统。

作品的中间贯穿着"时间之箭"，以黄浦江蜿蜒的形态为蓝本，象征着一段抽象的时间轴，连接和叙述着过去、此刻以及未来。上海的历史、时光流转在黄浦江上，如一只只轻舟过隙，如梦如世。

这件以艺术的方式来展现的船屋，希望为现实世界中的风景增添浪漫的想象和诗意。如中国古代文人将自我情怀寄托在画里的轻舟之上，这件作品也同时在传递一种符合当代的文化与艺术的情怀。

Boat expresses strong symbolism in both Eastern and Western culture. In Buddhism, it is an ark that carries human suffering to a peaceful world; in Taoism, it is a metaphor for renouncing the secular. In Western religions, there is a story about an ark saving all the creatures in the world. While in the artist's mind, boat embodies a strength which is warm yet firm as a mother's embrace, taking us on a journey in search of a peaceful shore, a spiritual home. The artist aims to create a boat-shaped sculpture, "floating" on the Huangpu Riverside. Standing by side of this busy river, it will create a contrast between motion and stillness, the imaginary and real space.

The upper part of the work is a boat house which represents a contemporary architectural structure in abstract form. Through this boat house, viewers can recall and imagine the old buildings in the docks of the Huangpu River of the past. Walking around the sculpture, one can see different dimensions of the structure from different angles. The sculpture therefore provides an alternative perspective to look at architecture, culture and tradition.

The middle part is in a linear form based on the shape of the Huangpu River. Symbolizing an "arrow of time", it connects the past, the present and the future of the river. The River carries the history and stories of Shanghai. As boats come and go on the river, scenes of everyday life seem real, yet they float by as if in a dream.

This boat house is presented in the form of art, which the artist hopes to add a poetic touch to the landscape of the real world. As ancient Chinese scholars often expressed themselves by painting a small boat in their works, this sculpture, too, presents a contemporary aesthetic in the modern age.

13

Irresolute

徊

沈烈毅
Shen Lieyi

类别：雕塑类艺术品
创作年份：2019
尺寸：5.6米×1.6米×0.99米
米
材质：不锈钢、山西黑花岗石
地点：煤气厂公园东

Type: Sculpture
Year of Creation: 2019
Size: 5.6m×1.6m×0.99m
Material: Stainless steel,
Shanxi black granite
Location: Gas Factory Park
East

上海是中国最早使用煤气的城市之一。杨树浦煤气厂于1933年建成投产，号称远东第一"自来火房"。1999年下半年起，因产业结构调整，杨树浦煤气厂停止生产。如今该区域为公共绿地，其北侧将作商业开发。

该项目位于煤气厂的堆煤场旧址上，场地被分为长60米、宽18米和长80米、宽18米的两块，地面为草坪。

空缺的椅子如同逝去的记忆，与月光同来，在如镜般的江面留下痕迹；倘若有人在此相对而坐，他不仅是与文化经验的一部分相遇，也是在与自身、他人相遇，而河流，将裹挟这一切永久地流淌。

Shanghai was one of the first cities of China which utilized coal. Yang Shu Pu Gashouse was founded in 1933 and put into use in the same year. It was called East Asia's No.1 firehouse. Since the second half of 1999, due to industrial restructuring, it was closed. Now the field is public lawn, and its north side is planned to be commercial development.

It is located on the coal-piled site of the old gashouse. There are two sites that one is 60 m long and 18 m wide, and the other ls 60 m long and 18 m wide. The ground is lawn.

Vacant chairs, like lost memories come with the moonlight, leaving traces on the river like a mirror; if someone sits here, he not only meets part of the cultural experience, but also with himself and others. The river will wrap it all forever to flow.

沈烈毅
1969年生于杭州，1995年毕业于中国美术学院雕塑系，现生活、工作于杭州。沈烈毅现任中国雕塑学会常务理事、浙江省雕塑研究会副会长，任教于中国美术学院雕塑与公共艺术学院，在中国及海外举办展览并参加艺术节、大奖赛等。他在中国美术学院任教的同时，也积极地开展创作活动。艺术家擅长使用岩石、金属等坚硬素材创作以湖面、雨水等为主题的雕塑，将两种相反的元素融为一体，创作出仿佛将水滴落下的一瞬间冻结一般的作品。

Shen Lieyi
Born in Hangzhou in 1969.Graduated from the Sculpture Department of China Academy of Art in 1995.Lives and works in Hangzhou. Shen Lieyi is currently the executive director of the Chinese Sculpture Society, and the vice chairman of the Zhejiang Sculpture Research Society. He also teaches at the Sculpture and Public Art College of China Academy of Art, and has organized exhibitions in China and overseas as well as participated in art festivals and awards. While teaching at the China Academy of Art, he actively initiated several creative activities. He is professed at using rock, metal and other hard materials to create sculptures based on the idea of lakes and rain, combining two opposite elements into one and creating works that seem to freeze instantly when the water drops fall.

14

黄浦货舱
Huangpu Hold

理查德·威尔逊
Richard Wilson

类别：雕塑类艺术品
创作年份：2019
尺寸：3米 ×5米 ×3米
材质：切割船舶型材、钢材、
工业涂料
地点：浮码头空地

Type: Sculpture
Year of Creation: 2019
Size: 3m×5m×3m
Material: Cutting ship
profiles, steel, industrial
coatings
Location: Floating Pler
Square

这件作品是关于从过去的河流中提取一个重要的工业物体，它们被丢弃，成为一个碍眼的东西。这一重要的物件已得到更新、改造并再次赋予生命，以提醒人们记住这条河流的强大历史。

关于封口切片的设想，艺术家取材于货轮的钢制外壳（船体、舱板和船的上半部结构），然后用焊炬切割钢板或者处理实心钢管（如扶手等），根据具体开口位置与角度制作成圆形或者椭圆形的切片。

在进行钢板切割之前，需要制作一个标记轮廓的工具。艺术家将一支激光笔绑在一片直径约6米的圆形木片的边缘，使其可以像自行车轮一样转动。接着，将这个标记工具固定在一个支架上，与地面垂直并正对着需要切割的钢板。当这个圆片旋转的时候，投射在钢板上的激光笔光点也随之旋转，最后连接成一个完美的圆形。

这时候需要有人用白粉笔沿着这个光圈描下轮廓，最后得到的就是一个精准的圆圈，而工人就可以根据这个图样切割了。

每一片圆形或者椭圆形的切面都要与钢管的末端截面焊接在一起，数个这样封口的管子堆叠在一起即形成这件雕塑作品。钢管的末端必须与这些切片精确吻合，因此在截取钢管时需要采取相同的角度。正如之前所述，只有激光笔标记工具与地面呈90°角，才可以在钢板表面得到一个完美的圆形轮廓，与钢管的圆形截面相契合。换言之，如果激光笔标记工具与地面呈其他角度，比如说45°角，那么得到的就是一个直径6米的椭圆轮廓，切割下来的椭圆切片覆盖的就应该是截面是45°角的钢管末端。

这些钢板切面的形状尽量各不相同，圆形、椭圆形，甚至是方形都有（如果钢管为方形开口），然后焊接在角度契合的钢管截面上。制作切面的金属板可以来自船体外壳的各种部分，包括窗户、边缘、扶手、甲板、管道、配件、水龙头等。所有这些部分制作的切面都可以焊接在任意管道的末端，但是尽量随意排列，以营造出一种拼接混搭的美感。最后呈现的作品就是一堆直径6米的钢结构管道，每根末端是角度不同的横截开口，并由取材自船体外部钢板的形状相同、边缘契合切片焊接封口。

The work is about taking an important industrial object from the rivers past that was discarded and thrown away to become an eyesore. This important object has been renewed, transformed and given a life once more as a reminder of the rivers powerful history.

What artist visualises is that the ships outer steel shell plate...(being the hull, decks and superstructure) are to be "cored" or cut with a flame torch. These cuts are all circles or ellipses depending on where the cut is made.

A marking out tool will need to be made to be able to mark out each hole. The artist uses a laser pointer and attaches this at the edge of a moving wooden circle approx.600 mm diameter and it must be able to turn (like a bicycle wheel). If this is on a stand and is positioned straight at the hull or deck or cabin it will describe as a moving laser dot... a perfect circle on the steel surface as it is turned round.

As the laser dot moves round it will need somebody to draw in white chalk where the laser has been tracked. The result will be a precise circle marked onto the ship that can then be followed to cut out using a flame torch.

Each circle or ellipse cut out will then be welded onto the end...(Each end) of a pipe that forms a stack as the sculpture. The ends of the pipes will be cut to the same angle as the circles or ellipses making a snug and accurate fit. If the circular tool holding the laser is set at an 90-degree angle, (right angle) to the steel it describes a perfect circle. However if it is set at any other angle say 45 degrees to the steel ship detail, it will give a 600mm ovoid or ellipse shape that can still be cut out and welded to a 45-degree cut on the end of the pipe.

The idea is to cut as many 600 mm varying ovoids and circles to fit onto each end of a number of pipes that will then be stacked up. Some pipe ends will be square cut. Other pipes will have differing angle cuts made to allow the ovoid to be welded neatly on. The idea is to draw the circles and ovoids onto different details or part details of the outer steel shape of the ship including bits of window, edges, hand rail, deck, pipes, fittings, taps etc. All these details are located onto either end of the pipe stack but can be arranged randomly so that it is not just the hull on the bottom row of pipes but includes other bits of the ship as an amalgamate. The final sculpture will be a stack of 600 mm diameter pipes all with differing angled cut ends that have welded onto them the many various elements or part details of the complete ship steel exterior.

理查德·威尔逊

1953 年生于伦敦，美国著名的雕塑家和装置艺术家。其作品大胆介入建筑空间，富有革新性。艺术家从工学和建筑学中汲取了大量的灵感。威尔逊拥有 30 年以上的国际参展经历，在美国、日本、巴西、墨西哥、俄罗斯、澳大利亚以及欧洲各国的主要美术馆举办过展览，并创作作为公共艺术的作品。曾经参加 1992 年悉尼、1989 年圣保罗双年展，横滨三年展以及越后妻有大地艺术节等。

Richard Wilson

Born in 1953.One of Britain's most reknowned sculptors and art installation artists. He is internationally celebrated for his interventions in architectural space which draw heavily for their inspiration from the world of engineering and construction. Wilson has exhibited widely, nationally and internationally for over thirty years and has made major museum exhibitions and public works in countries as diverse as Japan, USA, Brazil, Mexico, Russia, Australia and numerous countries throughout Europe, Wilson was also invited to participate in the Sydney (1992), Sao Paulo (1989), Venice and Aperto Blennial and the Yokohama Triennal, Echigo-Tsumari Art Triennial (2000) in Japan, and so on.

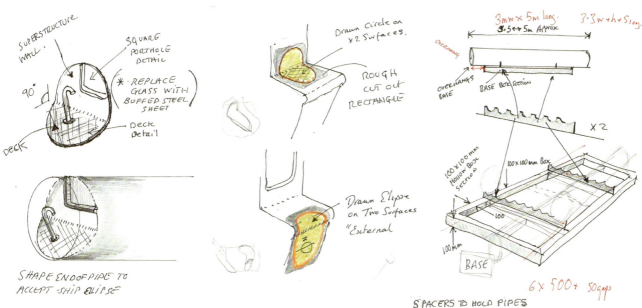

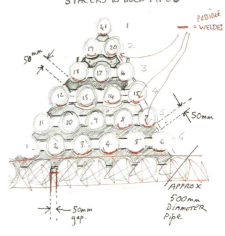

理查德·威尔逊：通过对废弃船只进行取材、切割、清洁，创造出和以前完全不一样的东西，以此提示它的过去、它的历史，以及水、交通运输和流动性的概念。我选择造船的材料，用船自己的语言来呈现船的作品。我要用雕塑让人们换一个视角看世界。人们不仅有视觉，也有听觉、嗅觉、味觉。在这个场地上，雕塑能帮助人聚焦，只要转身看看四周城市里的景色，心就会静下来。此时，人们回到雕塑，回到景观，回到场景。

空间及其周遭事物、锈迹斑斑的工业建筑、新与旧、寄身于当下的历史、黯淡的背景和原因，还有这艘船从前的任务——把废料拉走、再把原料拉回来，都是我的信息来源。我希望我的作品能与这个地方过去所存在的工业遗迹产生共鸣，希望人们能做出某种回应，认识一段历史。

Richard Wilson: Through sourcing, cutting and cleaning discarded ships, I create something completely different from what they used to be, reminding of their past, history, and the concept of water, transportation and mobility. I choose to use the materials of ship and present the artwork of ship in its own language. I want to use sculpture to make people see the world from a different perspective. People can not only see, but also hear, smell and taste. On this site, sculpture can help people focus, as long as you turn around and look at the surrounding scenery of the city, you will calm down. At this point, people return to the sculpture, to the landscape, to the scene.

These are all my resources of information: the space and its surroundings, the rusting industrial buildings, the old and the new, the history of the present, the bleak context and reasons, and the ship's former mission -- to haul away waste and bring back raw materials. I want my work to resonate with the industrial relics on the site, to make people respond in some way and perceive a piece of history.

15

『目』之廊
Gallery of "Mé"

目 [mé]
目 [mé]

创作年份：2019—2020
尺寸：
《丙烯酸气体大》：直径 2.3 米，厚 0.42 米；
《丙烯酸气体小》：直径 1.7 米，厚 0.18 米；
《接触》：3.6 米（宽）×8.5 米（高）；
《线》：约 6 米（宽）×2 米（高）；
《分离物》：2.64 米（长）×1.8 米（宽）×
1.96 米（高）；
沙发：1.94 米（长）×0.93 米（宽）×0.55 米（高）
材质：环氧树脂、木材、FRP、铜、发泡海绵等
地点：攀岩场

Year of Creation: 2019-2020
Size:
Acrylic Gas Big: d=2.3m, THK=0.42m;
Acrylic Gas Small: d=1.7m, THK=0.18m;
Contact: 3.6(W)m×8.5(H)m;
Line: about 6.0(W)m×2.0(H)m;
Respective Object: 2.64(L)m×1.8(W)m×1.96(H)m;
The sofa: 1.94(L)m×0.93(W)m×0.55(H)m
Material: Epoxy resin, wood, FRP, copper, foam sponge, etc.
Location: Climbing Gym

目 [mé]

日本艺术小组 / 团队，三位核心成员分别是艺术家荒神明香、艺术监督南川贤二和艺术制作经理增井宏文。目 [mé] 创作的艺术品操纵着对现实世界的理解。他们的装置激起了对周围世界内在的不可靠性和不确定性的意识。2013 年濑户内国际艺术节发表作品《迷宫之街～变化莫测的小路空间～》，2014 年为宇都宫美术馆馆外项目创作作品《有一天，一个阿伯的脸飘在天空中》，2017 年参加北阿尔卑斯国际艺术节创作《信浓大町实景舍》等，使用各种方法创作各种类型的作品。

目 [mé]

A Japanese art collective/team with three core members-artist Haruka Kojin, director Kenji Minamigawa, and production manager Hirofumi Masui. Mé works on the realization of artworks that manipulate perceptions of the physical world. Their installations provoke awareness of the inherent unreliability and uncertainty in the world around us. In 2013, they published work *Maze Street - Unpredictable Pathway Space -* in the Setouchi International Art Festival; in 2014, they created *The Day an Ojisan's Face Floated in the Sky* for the Utsunomiya Museum of Art outdoor project; in 2017, they participated in the Northern Alps International Art Festival and created *Shinano Omachi Real View House*, etc., and uses diverse methods to create various types of works..

攀岩场点位位于杨浦滨江南段北侧，上海工部局电气处新厂旧址内，北邻上海国际时尚中心，南邻上海煤气公司杨浦工场旧址。上海工部局电气处新厂是一处重要的近现代工业遗产，筹建于 1910 年，1913 年 4 月 12 日正式发电，1929 年更名为上海电力公司。2010 年 12 月 18 日正式停止运行。

在大规模的城市再开发中不断改变的上海，仿佛一场远远超过个人的行动范畴的巨大的运动，在我们的面前发生、上演。然而，我们的"眼睛"却总是被悬挂在上海空中摇晃着的晾干的衣物，或是年头已久的巨大烟囱上的锈斑，又或是在新起的建筑物与废墟间自由来去的小鸟的细微的举动所吸引。景观不断美丽地重生、再造，在这样的巨大能量之下，艺术家希望有一个和个人的视线共鸣的场所。艺术家希望创造一个，让来访的每个人的各自的视角、观点都生机勃勃的空间。

这将是一个催生出每一位观赏者各自的"视角"的作品，由数个作品构成的如美术馆般的寂静空间。

Climbing Gym is located in the north of The South Binjiang Section of Yangpu, in the old site of the electrical Department of Shanghai Ministry of Industry, near Shanghai Fashion Centre in the north and Yangpu Workshop of Shanghai Gas Company in the south. The new factory of electrical Department of Shanghai Bureau of Industry and Technology is an important modern industrial heritage, which was founded in 1910 and generated electricity at 12 April 1913. In 1929, it was renamed Shanghai Electric Power Company. It officially ceased operation on December 18th, 2010.

Shanghai is changing in the midst of massive redevelopment, as if a huge movement far beyond the scope of individual action is taking place in front of us. But our "eyes" are always drawn by the swaying dry clothes hanging in the Shanghai air, or the rust stains on the old giant chimneys, or the subtle movements of the birds that move freely among the newly built buildings and ruins. The landscape is constantly being reborn and recreated in a beautiful way. With such great energy, artist hope to have a place that resonates with my personal vision. Artist wanted to create a space where everyone's perspective and view would come alive.

It will be a work that brings out each viewer's own "perspective", a silent space composed of several works like an art gallery.

16

起重机的对角线
Set of Diagonals for Cranes

费利斯·瓦里尼
Felice Varini

类别：绘画类艺术品
创作年份：2019
尺寸：22米×29米
材质：油漆
地点：吊塔

Type: Painting
Year of Creation: 2019
Size: 22m×29m
Material: Paint
Location: Tower Crane

费利斯·瓦里尼：第一次来到杨浦滨江，我看到了这个巨大的、不再使用的起重机。这些橙色的大型结构装置，在某种程度上，就像是一个代表过去工业历史的考古对象，向我们传达了已经消失的一段时期、一段故事。

在这件作品中，我建立了一个双重交叉的视角——两组交叉线，向四个方向，以特定方式无限向外散射，与它们所接触的所有元素相遇。我希望通过起重机的巨大体量以及视觉的深度，产生出极具冲击力的效果，也赋予已停止使用的起重机以跃动感，重现其过去曾有的力量感。

起重机是橙色的，我只用白色作为补充和对比。我不希望作品中再出现其他色彩，毕竟生活中色彩无处不在。在杨浦滨江，河有其颜色，天空有其颜色，这些色彩随着户外光线的变化，也一直在变化着。

我的作品其实是一幅绘画作品，我总是根据新的场地，画出我的画作。

Felice Varini: The first time I came to Yangpu Waterfront, I saw this huge crane which is out of use. These large orange structural installations, in a way, are like an archaeological object representing the past industrial history, telling us a period, a story that has disappeared. In this artwork, I have created a double-crossing perspective—two sets of crossing lines that, in four directions, scatter infinitely outward in a specific way, encountering with all the elements they touch. I hope that through the huge volume of the crane and the depth of vision, to produce a very powerful effect, but also to refill the crane with dynamics, to represent its past sense of power. The crane is orange and I only use white colour as complement and contrast. I don't want other colours to appear in my work. After all, our life is colourful. At Yangpu Waterfront, the river has its own colour, the sky has its own colour, these colours are always changing with the change of outdoor lighting. My work is actually a piece of painting, and I always paint my paintings according to the new site.

艺术家创作的场所可以是建筑空间，也可以是构成类似空间的任何物体——它们都是其绘画的原始媒介。艺术家每次在不同的空间"现场"创作，作品从这个空间生长出来——《起重机的对角线》就是在杨浦滨江的吊塔上现场创作的作品。这件作品取形于三座橙色塔吊。此处，艺术家希望扰乱起重机的次序，来缅怀过去的时光。从东侧望向起重机，可以看见几何构图，这些图形使它们彼此共振。而散步的观看者将感受到这些用金属结构进行绘画的空间变体。

The place the artist creates can be an architectural space or any objects that make up space-they are the original medium of his paintings. Every time the artist creates "on-site" in a different space, the work grows out of the space. *Set of Diagonals for Cranes* is a work created on-site on the crane towers along Yangpu Waterfront. The work takes shape on a set of three orange cranes. Here the artist's wish is to play with this set of out of order unloading cranes, memory of a past time. It is a matter of putting them in resonance with each other, by a geometric composition visible from the point facing all the cranes on the east side. The stroller viewer will see the multitude of spatial variants of painting playing with these metallic structures.

费利斯 · 瓦里尼
1952 年生于瑞士，现居巴黎。艺术家擅长将包括室外、室内在内的整个空间作为画布创作大型作品。他使用投影仪将几何图形投影并绘制在建筑物的墙壁和梁柱上，从特定地点看过去，立体的建筑物上便会显现出平面的图形。瓦里尼的作品存在于世界多地的公共空间,给日常生活中平凡的场景以新鲜的体验。

Felice Varini
Born in Switzerland in 1952.Currently resides in Paris, France. Varini turns interior and exterior space into large works of art. Using projectors, he projects geometrical shapes on walls and pillars, traces them, and standing from a certain point, it can seem like a whole building is rendered flat with his geometric design. By doing such. he takes things from everyday scenery and allows us to experience a new world. He has created such works in public spaces all over the world.

17-1

诗人之屋
A Casa Dos Poetas (The House of Poets)

荷塞·德·吉马良斯
José de Guimarães

该项目位于"灰仓艺术空间"最东侧筒仓内部，筒仓半径约 7 米、高约 1 米。灰仓美术馆由原电厂粉煤灰储灰罐改造而来，作为 2019 城市空间艺术季的分场馆使用，其内部螺旋坡道为钢结构，具有较好的观赏条件。

灰仓艺术空间位于"拱门"所在步道的附近。这座颇有意味的建筑由三个巨大的煤灰仓库组成，它们曾属于一座历史悠久的工业电厂。三个圆柱形构筑物中的一个，内部完全上下打通，形成一个高达 13 米的空间。

艺术家决定把这一巨大的开放空间转变成一个朦胧地带，在这里安装一组彩色霓虹灯——分别是红色、黄色和蓝色，它们以 10 秒的间隔依次闪烁。霓虹灯管再现了拱门横梁的上侧曲线，从而与安置于黄浦江岸的艺术品相联结。吉马良斯将其取名为"诗人之屋"，但它也可以称为"冥想之屋"。这导向了数个疑问。

人类为什么聚集起来进行冥想？为什么选择聚集在一所房屋中？这座"灰仓美术馆"，或更贴切来说，"诗人之屋"如同一座寺庙，此间恒久寂静。这是一座个人内在思索与安静避世的理想之居。霓虹灯不仅仅是一种装饰，更在文化交融之处营造一种创造性的环境。这不仅是一个对话与交流的空间，也是不同人群和文化发生真正交融的空间，即，不同的文化在这里相遇，揭示世界和谐的潜能。"诗人之屋"是一处人们聚集在一起聆听内在的安静居所，不同文化通过黄浦江所缔造的道路来到此地，引发这种和谐或"文化渗透"，将世界各地的文化联结在一起。

The project is located inside the easternmost silo of the Ash Bucket Art Space, with a radius of about 7 m and a height of about 13 m. The Ash Bucket Museum is transformed from the original power plant's fly ash storage tank. It was used as a sub-venue of the 2019 Urban Space Art Season. The internal spiral ramp is a steel structure and has good viewing conditions.

This cylinder building is near the road where the artist's work, Gates, is located. It's an interesting building consisting of three enormous cylinders that are part of a historic industrial plant. One of the three circular buildings is 13 m high and its interior is a totally open space.

It was decided to transform this large open space into a twilight zone, where a set of colourful neon lights-red, yellow and blue-were installed and flash sequentially at 10-second intervals. These neon tubes represent the line of sight of the upper part of the gates, thus creating a connection with those works of art built on the Huangpu River margin. Guimarães called this cylinder "Casa dos Poetas" (House of Poets), but it can also be called "Casa de Meditação" (Meditation House). This raises various questions.

Why do human beings gather to meditate? Why do they choose a house to gather? This "Cylinder Building" or, more fittingly, "Casa dos Poetas" is like a temple, where silence predominates. It is an ideal place for personal inner thoughts and for a peaceful retreat. The neon lights are not merely an adornment or a decorative piece but, rather, help to generate a creative ambience where cultures mingle. It's a space, not so much for dialogue, or for mere communication, but for authentic communion between the various peoples and cultures, that is, a place where cultures encounter one another and reveal the potential of universal harmony. It is this harmony or "cultural osmosis" that "Casa dos Poetas" emanates, a place of silence where persons gather to listen to their inner voice and where different cultures come together through routes and paths like the one the "Huangpu River" forged to bring all the world's cultures together.

创作年份：2019
尺寸：13.8 米（高）×6.8 米（宽）
材质：霓虹灯彩管、钢缆
地点：灰仓艺术空间

Year of Creation: 2019
Size: 13.8(H)m×6.8(W)m
Material: Neon tube, cable
Location: Ash Bucket Art Space

17-2

A World
一个世界
高桥启祐
Keisuke Takahashi

高桥启祐

曾在意大利米兰、中国台北的画廊，以及日本 BankArt1929 等机构举办个展。他曾参加上海双年展（2004）、濑户内艺术三年展（2016）、雅加达双年展（2017）等大型国际展览。2005 年，高桥在日本"第九届文化厅媒体艺术节"获得"学委会推荐优秀作品"奖。他在舞蹈团体"Nibroll"担任视觉总监，创作了多部表演艺术作品。高桥启祐注重探索身体与影像、空间之间的关系。他将"身体"作为创作主题，通过这层滤镜来重新审视社会。他试图重新审视因不同空间而产生的差异化个体存在，并通过主题进行实验，以达成人与人之间的沟通。个体既存在于公共空间中，也存在于私人空间中。高桥启祐通过思考个体存在如何与群体相互连接，进而探索身体与空间的关系。

Keisuke Takahashi

He held the solo exhibition in the gallery in Milan, Italy and Taipel, China, BankArt1929 in Japan and others. He also jointed to the international exhibition such as Shanghai Biennial (2004), Setouchi Art Triennial (2016), Jakarta Biennial (2017) and others. He received the Committee Recommendation Award from Japan Media Arts Festival in 2005.He also created many performance works as the visual director for the dance company named Nibroll. Keisuke Takahashi always think about the relationship between the body, image, and the space. He creates works by the theme of a body, reconsidering the society through the filter of the body. His purpose is to re-consider the existence of an individual as what comes from a difference caused by spaces, and place the theme in a trial to develop it into communication among human beings. An individual exists in both public and private places Keisuke Takahashi think about how personal existence is connected to the community to consider a relationship of a body and the space.

成百上千的人在屏幕之间游走。这些无数的人形会在最终汇聚成世界地图。我们迷失了方向，四处游荡。不和、摩擦和冲突。这件作品描述了现今不沟通的模式，以及对价值观的一种希望，这价值观是反复无常的、无可计数的，但可能是相互联系的。

这件作品由各种人们行走的剪影组成，艺术家让他们环游了世界。美洲人形成了美洲，欧洲人形成了欧洲，亚洲人形成了亚洲，而各个国家的人民组成了一个世界。人们的行走方式似乎因环境、地点、种族或宗教而有所不同。

我们眼前的世界处于一个不同文化、不同价值观的人们以多样的形式共生的时代，但世界既简单又复杂。"共生"不仅包括和谐和融合的意义，也包括妥协、冲突、矛盾和复杂的意义。

新世界将会在另外一端。

Hundreds and thousands of bodies go across some screens. Eventually,those countless bodies turn into the shape of a world map. We lose our direction and our ways and wander around. Discord, friction and conflicts. The work describes the present form of discommunication and a hope about our values inconsistent and countless but possibly connected with one another.

This work is made of walking silhouettes of various people. Artist took them around the world. Americans are forming Americas, Europeans as Europe, Asia as Asia, people in their respective countries form one world. People's walking seems to be different depending on the environment, location, race or perhaps religion.

The world in front of our eyes is an age where people with diverse cultures and diverse values live together in diverse directions. But the world is both simple and complex. "Symbiosis" includes not only the meaning of harmony and fusion, but also the meaning of compromise, conflict, contradiction, and complex.

The new world will be on the other side.

类别：装置艺术品
创作年份：2019
尺寸：
投影面：3.48 米（高）×38.47 米（长）；
空间：高 4.48 米，直径 13.6 米
材质：360° 环绕投影
地点：灰仓艺术空间

Type: Installation
Year of Creation: 2019
Size:
Projection surface: 3.48(H)m×38.47(L)m;
Space: The height is 4.48 m, and the diameter is 13.6 m
Material: 360° surrounded projection
Location: Ash Bucket Art Space

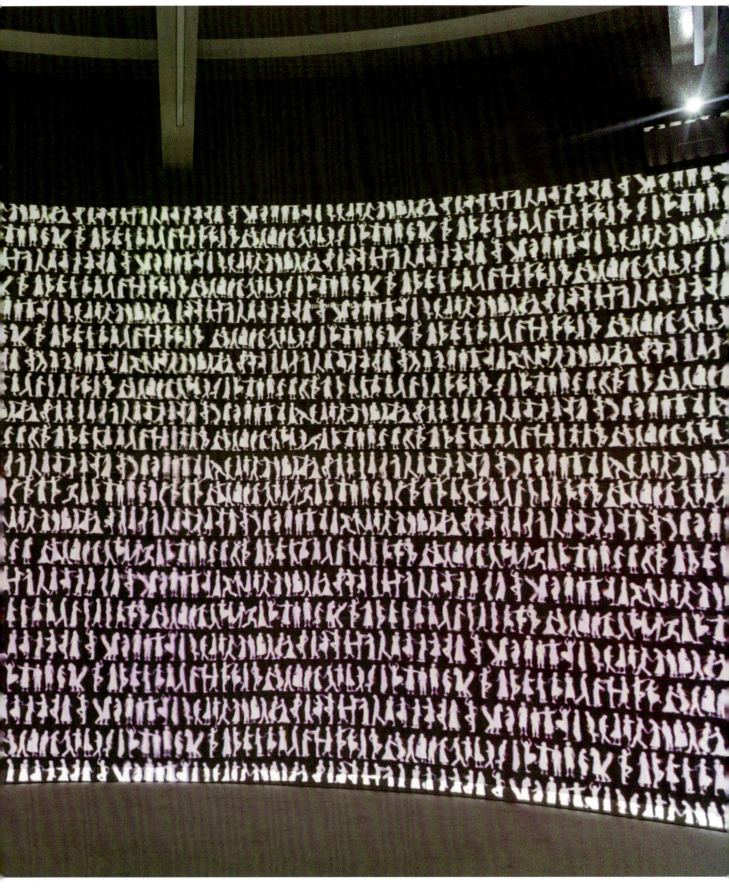

17-3

Encounter 相遇

韩家英 Han Jiaying

韩家英

1961 年生于天津。1986 年毕业于西安美术学院。1993 年创立韩家英设计公司。国际平面设计联盟（AGI）会员，中央美术学院城市设计学院客座教授。2003 年在法国举办《天涯》专题设计个展；2012 年首届中国设计大展平面策展人；2012 - 2014 年在京、沪、深三地举办"镜像·韩家英设计展"；2014 年第一届深港设计双年展视觉创意总监；2016 年德国红点设计大奖评审；2016 年受邀参加华沙海报双年展五十周年特别展"50/50/50"，曾荣获亚洲最具影响力设计大奖金奖，福布斯 2015 中国最具影响力的设计师。多项优秀作品收藏于包括 V & A 博物馆在内的英国、法国、德国、丹麦、日本、中国等国际艺术机构。

Han Jiaying

Born in Tianjin in 1961. Graduated from Xi'an Academy of Fine Arts in 1986. Established Han Jiaying Design Company in 1993. Member of the Alliance Graphique Internationale (AGI). Visiting Professor, School of Urban Design, Central Academy of Fine Arts. Held a solo exhibition of "The End of the World" in France in 2003. Plan Curator of the first China Design Exhibition 2012. 2012-2014 held the "Mirror · Han Jiaying Design Exhibition" at the CAFA Art Museum, OCT Art & Design Gallery, and Shanghai ROCKBUND. In 2016, he was invited to participate in the special exhibition "50/50/50" of the 50th Anniversary of the Warsaw Poster Biennale. Won the DFA Design for Asia Awards. Forbes 2015 China's most influential designer. Works are collected by many international art institutions in the United Kingdom, France, Germany, Denmark, Japan, China, including V&A Museum.

作品《相遇》由 12 组方、圆等不同的形体组合而成，试图探索精神上、物理上、时空中，关于相遇的无限可能。作品由镜面、石材、木、金属等材质来演绎，让观者沉浸"形 / 色"交融之间，形成艺术投射式观展体验，象征人、艺术与世界的多元相遇，寓意滨江视界与艺术境界的共生、共见，将带来全新的美学启迪与艺术启发。

艺术家在创作过程中，完全将"相遇"这个主题作为思考、创作的问题。在这个基础上，艺术家不断地去找各种材料，在找寻"相遇"感觉的过程中，思路越来越清晰，也越来越具体，使观看的人在作品中看到材料之间的对话、材料与观者的对话，产生出多个层面的"相遇"。

The work *Encounter* is a group of 12 works composed of different shapes such as squares and circles. It tries to explore the infinite possibilities of encounters spiritually, physically, in time and space. The work is interpreted by mirror, stone, wood, metal and other materials, so that the viewer is immersed in the blend of "shape/colour", forming an art projected viewing experience, symbolizing the multiple encounters of people, art, and the world, meaning the riverside vision and Symbiosis and common understanding in the art realm, which will bring new aesthetic inspiration and artistic inspiration.

During the creation, the artist put aside the thinking of design and completely regarded the theme of "encounter" as a question of thinking and creation. On this basis, the artist is constantly searching for various materials, and in the process of searching for the feeling of "encounter", his ideas become clearer and more specific. In the work, the viewers can see the dialogue between the materials, the dialogue between the materials and the viewers, and produce multiple levels of "encounter".

类别：装置艺术品
创作年份：2019
尺寸：大小不一
材质：钢板、镜面不锈钢、亚克力、大理石等石材、玻璃、原木、原石
地点：灰仓艺术空间

Type: Installation
Year of Creation: 2019
Size: Varies in size
Material: Steel plate, mirror stainless steel, acrylic, marble, stone, glass, logs, rough stone
Location: Ash Bucket Art Space

17-4

钢之迷宫 & 灰之迷宫
Maze of Steel & Maze of Ash

章明
Zhang Ming

章明
同济大学建筑与城市规划学院建筑系副主任、教授、博导；同济大学建筑设计研究院（集团）有限公司原作设计工作室主持建筑师；上海市建筑学工学术委员会学术委员、建筑评论学术委员会委员、小城镇建筑分会副会长；《城市环境设计》《建筑技艺》《建筑实践》等杂志编委。

Zhang Ming
Professor/Vice Head of Department of Architecture, College of Architecture and Urban Planning, Tongji University, Directing Archtect, Original Design Studio, TJAD, Executive director of the Shanghai Architectural Society, director of the architectural creation academic department, Vice President of the Small Town Construction Branch, Editorial Member of *UED*, *Architectural Technique* and *Architectural Practice*.

类别：装置艺术品
创作年份：2019—2020
尺寸：各高 4.48 米，直径 13.6 米
材质：钢板、造雾机、激光灯
地点：灰仓艺术空间

Type: Installation
Year of Creation: 2019-2020
Size: For each of the mazes, the height is 4.48 m and the diameter is 13.6 m
Material: Steel plate, fogger, laser light
Location: Ash Bucket Art Space

灰之迷宫营造出筒仓内部原本的灰雾蒸腾的意向，与钢之迷宫路径对应，限定路径的介质由钢板变成虚化的镭射光线。钢之迷宫取废弃板壁搭建出迷宫，勾勒出男女不同的探寻路径。人们在路径穿梭中与筒仓的历史与现状相遇，人与人之间相遇相离。

Maze of Ash is going to create the original intention of grey mist transpiration inside the silo, corresponding to the labyrinth path of steel. The medium that defines the path changes from steel plate to virtual laser light. While the Maze of Steel builds a maze out of the abandoned board walls, and outlines the different search paths for men and women. People meet with the silo's history and current situation in the path shuttle, and people meet and leave each other.

规划建筑版块
Planning & Architecture

规划建筑版块生动演绎了国内外滨水空间的规划建设理念，综合呈现上海"一江一河"建设的成就和未来展望，丰富了本届空间艺术季的知识性和趣味性，激发人们更多的理性思考，引导形成对滨水空间发展目标的社会共识。

滨水空间为人类创造了美好生活。但水起到的作用并非"水到渠成"。如果水有任何魔力，则取决于"美"和"好"在人类生活中不可分割。苏格拉底式的"好"与孔夫子的"仁"不谋而合，皆为人类追求美轮美奂的生活赋予了道德的分量。水因其丰富的象征性及顽强的转化能量，成为人类美好生活创造的源泉与犀利的工具。人类营造滨水空间历史表明，"美好生活"，因为水的存在与作用，既非单纯享乐主义式的幸福生活，亦非现代人所谓生活方式的选择。

水之魔力通过"三城记"——上海的"一江一河"、威尼斯和悉尼——得以显现；滨水的故事与未来揭示水如何改变人的生活及城市空间。"水之魔力"其实即是人与生活的魔力；通过文化与地域的邂逅，我们以叙事（规划建筑案例）展现滨水空间给城市与人的生活带来再生。

This section vividly interprets the planning and construction concepts of waterfront spaces at home and abroad, comprehensively presents the achievements and future outlook of the construction of the area along the Huangpu River and Suzhou Creek, enriches the knowledge and interest of this year's SUSAS, stimulates more rational thinking, and leads to form the social consensus on the development goals of waterfront spaces.

Waterfront space has helped to provide humans with good and beautiful life experiences. But realizing such potential does not simply arise naturally. The magical power of water depends on the inseparability of beauty and the goodness. It is the Socratic concept of the good – the same as the Confucian idea of morality – that gives beautiful life its moral weight. Because of its rich symbolism and the resilient transformative capacity, water has become both an inspiration for and an effective tool of the humans in their pursuit of "the beautiful and good life". The history of waterfront space shows that, due to water's existence and its instrumentality, the good life is not merely a hedonistic happy lifestyle choice, as we moderns are inclined to believe.

Through the "tales of three cities" (Huangpu River and Suzhou Creek, Venice, Sydney), the magical power of water is revealed. The waterfront stories and future show how water change the lives of people and space in city. The enigma of water lies in the magic of life. Through the "encounters" of culture and place, we use narratives (urban and architectural projects) to reveal the rebirth and mutation of cities and of people brought about by the transformations of waterfront space.

P101

像素滨江
Pixelating Huangpu River

张海翱
Zhang Hai'ao

采用磨砂亚克力棒，底面的 LED 灯光将亚克力棒染成江川的色调，亦可通过 LED 的明亮与色调变化使该装置有动态的波动效果。

Frosted acrylic rod is adopted, the LED light on the bottom will dye the acrylic rod into the colour of jiangchuan, and the device can also have dynamic fluctuation effect through the brightness and colour change of LED.

悉尼是世界上最大的海港城市之一。320 公里长的海港前滨勾勒出一系列宽阔的海湾和深水入海口，涵盖了各种各样的自然景观和设计景观。入海口和外港中有裸露的悬崖、原始的海岸线和浓密的丛林景观，还有一系列受保护的海滩和海港边缘人迹罕至的滨水通道。由海湾和入海口组成的网络景观独特，无论是自然景观还是人工景观，都赋予了悉尼滨水公共空间独一无二的性格。作为城市发展的一部分，悉尼港前滨 50% 的区域都已经历重建。这一系列的图片、照片和影片探索了这一建筑前沿的细微变迁，展示了前滨是如何运作的，以及在不同的地方是如何通过不同的项目、以不同的方式重塑、重生和恢复的。本次展览中涵盖的范围很广，从古老的原住民居住地到大规模的新开发项目。这些地方和项目的可见度各不相同。有些需要仔细观察才能发现，而另一些则需要给予重视。这些建筑设计出的滨水空间主要风格是克制的、基于环境的，并对不同的社区功能使用非常上心。总之，这些地方揭示了悉尼独特的内在特点，更重要的是，这些设计持续一致地致力于保证前滨为公众所享，让悉尼的所有人都能共同享受海滨。

Sydney is one of the world's great harbour cities. The 320-kilometre harbour foreshore enfolds an array of broad bays and deep inlets across a diverse range of natural and developed landscapes. The exposed cliffs, raw coastline and dense bush landscape of the ocean entrance and outer harbour give way to a sequence of protected beaches and intimate waterfront pathways along the harbour's edge. The network of bays and inlets is marked with distinctive landscapes, both natural and constructed, which collectively give Sydney's waterfront public spaces a unique character. Fifty percent of the Sydney Harbour foreshore has been reconstructed as part of urban development. This collection of drawings, photographs and film explores the nuances of this constructed edge, showing how the foreshore operates and where it has been reshaped, reclaimed, and restored in different ways in different places and through projects. The places included in this exhibition range broadly from ancient sites of Aboriginal habitation through to large-scale new developments. The visibility of these places and projects varies. Some require careful observation to detect, while others demand attention. The prevailing language of these constructed waterfront spaces is restrained, contextual, and attentive to diverse community use. Together, these places reveal a unique and intrinsic Sydney character and, importantly, a sustained and coherent commitment to publicly accessible foreshore so that all the people of Sydney can share and enjoy the waterfront.

P201

流动
The Flow

华东建筑集团股份有限公司（张桦、沈迪、董艺、隋郁、顾一冰）、华建集团上海建筑设计研究院（陈文杰、刘勇、刘坤、林松、邢澄）、华建集团华东建筑设计研究总院（杨明、宿宸、袁野）、华建集团历史建筑保护设计研究院（宿新宝、苏萍、王天宇、姜旭）、华建集团上海市水利工程设计研究院（程松明、张丽芬、濮勋、陆扬）、华建集团上海建筑科创中心（沈迪、高文艳、李南、姜旭）、华建集团上海建筑设计研究院（王伟杰、卜骏、程怡冰、建集团建筑装饰环境设计研究院（朱冰颖、吴云超、朱兆一、倪若琦）、周展晨、刘明昊、华建集团规划建筑设计研究院（罗镔、莫霞、袁志豪、占琳、王慧莹、魏沅）

Arcplus Group PLC (Zhang Hua, Shen Di, Dong Yi, Sui Yu, Gu Yibing), Arcplus Institute of Shanghai Architectural Design & Research (Co.,Ltd) (Chen Wenjie, Liu Yong, Liu Kun, Lin Song, Xing Cheng), East China Architectural Design & Research Institute (ECADI) (Yang Ming, Su Chen, Yuan Ye), Arcplus Historic Building Conservation Design & Research Institute (Su Xinbao, Su Ping, Wang Tianyu), Arcplus Shanghai Innovation Centre for Building Science & Technology (Shen Di, Gao Wenyan, Li Nan, Jiang Xu), Shanghai Water Engineering Design and Research Institute Co.,Ltd. (Cheng Songming, Zhang Lifen, Pu Xun, Tian Liyong, Lu Yang), Arcplus Architectural Decoration & Landscape Design Research Institute Co., Ltd. (Wang Weijie, Bu Jun, Cheng Yibing, Zhou Zhanchen, Liu Minghao, Zhu Bingying, Wu Yunchao, Zhu Zhaoyi, Ni Ruoqi), Arcplus Urban Planning & Architectural Design Institute (Luo Bin, Mo Xia, Yuan Zhihao, Zhan Lin, Wang Huiying, Wei Yuan)

流动

"一江一河"记载着上海城市历史和文化,深刻烙印民众情感。从"一江一河"出发,引出本展区主题"流动"。流动既可以指水体的流动,也可引申为演变、传承。本展区通过历史的流动、人的流动和空间的流动三个层面,阐述"一江一河"的发展演变过程。本版块包含四个独立单元:"台口""水岸·口门""水岸相依·三道相随""一河两岸·网络相遇"。

第一部分
台口

今年的大主题是"相遇",河口公园是反映主题的最佳题材。这里不仅是上海城市江河碰撞的地理交汇处,也是上海登上历史舞台与世界目光交接的窗口。在苏州河的流动舞台上,城市的历史与未来穿越并置,集体记忆在此处以夺目的色彩缤纷呈现。

沙盘直径 2.4 米,描绘了以河口滨江两岸为主题的未来城市亲水生态。老建筑不变,是历史见证者;新环境巨变,代表未来的可能性。沿河聚集的引导性的蝴蝶是"苏州河的精灵",代表了人和时间的聚散交融。《台口》是真实世界的想象再现,是现实物质空间和未来时间碎片的重构。

第二部分
水·岸·口门

由反射金属象征河流串联起不同时期的水、岸、口门关系,通过河流、展台材料质地与高度的变化,辅以河道断面、口门图片、滚动相册等内容,展现不同历史时期"一江一河"在水、岸、口门等方面的变迁,以此体现出"一江一河"与上海市城市发展的关系。5 个展台喻示出"一江一河"发展进程中的 5个不同历史阶段,即 1840 年前、1840 年—1979 年、1979年—2000 年、2000 年—2019 年、2020 年后,使参观者按时

The Flow

"The River and the Creek (Huangpu River and Suzhou Creek)" note down Shanghai's history and culture and are engraved with people's affection. The exhibition theme "Flow" is then brought out by the River and the Creek. The flow could refer to the water, or further, the evolution and inheritance. The two's evolution process is showcased in the flow of three levels, history, people and the space. The exhibition has four independent sections: "Estuary", "Water & Revetment & Gate", "Better Life by the River" and "Cross-Straits of a River, Network Meets".

PART ONE

Estuary

The theme of this year is "encounter", where the estuary park exactly represents its spirit. This place is not only the geographical convergence of the Suzhou Creek and Huangpu River, but also a remarkable window for Shanghai to step on the historical stage and attract the world's attention. On the fluid platform of Suzhou Creek, the past and the future of urban space juxtapose and the collective memory of the place emerges, in a brilliant colour.

The model, 2.4 meters in diameter, describes the waterfront life of Suzhou creek of the futuristic urbans. The old buildings stand still, witness the history, while the new environment transform drastically, represents the future. The butterfly along the river is the leading elf of Suzhou creek, where the separation and reunion of people and time evolves. "Estuary" is the imaginary reappearance of a real word. It is the reconstruction of fragments of physical space and time, past and future.

PART TWO

Water & Revetment & Gate

The reflective metal symbolizes rivers and connects water, revetment and gate in different periods. Through the varieties of material texture and height in river and each booth, supplemented with the river sections, the gate pictures and the rolling albums, they show the changes of the water, revetment and gate in Huangpu River and Suzhou Creek, so as to reveal the relationships between rivers and urban development in Shanghai. The five booths have indicated five periods of the development in Huangpu river and the Suzhou creek, which are before 1840, 1840-1979, 1979-2000, 2000-2019 and after 2000. They let visitors experience the evolution of Huangpu river and the Suzhou creek chronologically. At the same time, different river section models in different periods are set up in each booth, in order to increase

间顺序体验"一江一河"的发展演变历程。与此同时，展台上还设置有可开启的不同历史时期的水闸门型，以增强与观展者的互动。

第三部分
水岸相依·三道相随

城市，因水而兴；空间，因水而灵动。黄浦江，上海的母亲河，哺育了独树一帜的海派文化，见证着上海的成长与发展。黄浦滨江贯通工程位于浦江西岸，北起外白渡桥，南至卢浦大桥，全长 38.3 公里。项目既是以三道（漫步道、跑步道、骑行道）沟通城市的滨水空间，又是展示工业文化、世博文化的大舞台。我们在这次展览中共同回顾，奔流不息的黄浦江，如何经过了百年风雨的洗礼，以更为开放宽容的姿态再次回到市民的怀抱。

第四部分
一河两岸·网络相遇

基于新静安区苏州河一河两岸地区城市设计中的思考，研究滨水地区的演变与城市地区发展之间的客观规律，从而提出规划引导的方法，探索城市设计中对公共空间的引导策略。

空间的相遇：突出自外而内的演变过程，建构网络化慢行体系，联系开放空间，强调滨水与腹地的活动连接与渗透。在功能业态方面，突出与腹地功能核心区的结构性联系，引入具有影响力、高能级的现代服务业和文化产业功能。

时间的相遇：积极挖掘有价值的人文历史点，注重新旧融合。增加高能级的文化艺术设施，塑造具有标志性的建筑和开放空间，体现人文沿承，展现文化魅力。

the interactivity between the exhibits and visitors.

PART THREE
Better Life by the River
The city prospers and the space turns vigorous because of the water. Huangpu River, the native river of Shanghai, has bred a unique Shanghai culture and witnessed the development of the city. The Huangpu Riverside Project is located on the west bank of the river, from the Waibaidu Bridge in the north, to the Lupu Bridge in the south, with a total length of 8.3 kilometers. The project connects the waterfront space of the city with three lanes (walking, running and cycling), and also serves as a stage to display the EXPO and industrial heritage. In this exhibition, we look back on this endless running Huangpu River, which was returned to the citizens with a more open and inclusive manner after hundred years of vicissitude.

PART FOUR
Cross-Straits of a River, Network Meets
Based on the consideration of urban design in the areas along Suzhou Creek and the First River in Xinjingan District, this exhibition studies the objective law between the evolution of waterfront areas and the development of urban areas, puts forward the method of planning guidance, explores the guiding strategies for public space in urban design, and achieves spatial connectivity.

The encounter of space: highlight the from-outside-to-inside evolution process, construct a network slow system, contact open space, emphasize the connection and penetration of activities along the waterfront and hinterland. In terms of functional art, highlight structural links with the core area of the hinterland, introduce an influential, high-energy modern service industry and cultural industry.

The encounter of time: actively explore valuable human history points, pay attention to fuse of new and old. Increasing high-level cultural art facilities, shaping an iconic building and open space, reflecting humanity, showing cultural charm.

P301

刘恩芳、范文莉
Liu Enfang,
Fan Wenli

日常的力量——十一个城市
设计策略点亮滨水日常生活
The Power of Daily Life — 11 Urban Design Strategies to Light up Waterfront Daily Life

刘恩芳、范文莉，两位设计师将"好的城市设计"视为是在物质空间层面支持并表达城市精神（非物质性目标）的重要工具和必备要素。以"开放性、互联性、可持续性、流动性和人本性"为一以贯之的设计哲学，从心出发塑造城市公共生活的日常力量。

秉持"Urban+"的突破思维与方法、设计策略及实施把控，刘恩芳、范文莉团队完成了雄安建筑风貌导则、上海虹桥高铁枢纽商务核心区（一期）城市设计、上海世博B片区央企总部基地城市设计等一系列广受好评的设计创作。

我们参展的主题为"日常的力量"。"日常的力量"是我们多年来持续关注城市公共空间品质提升的重要内容。我们通过城市设计，在更广泛的范围服务于市民日常生活，让城市公共空间与城市生活紧密相连，让城市空间品质提升成为更多人的福祉。本次展览旨在唤起人们对城市公共空间的主动关注，结合这次滨水主题，收集人们对城市日常生活的感受和需求，挖掘滨水空间与日常生活的关联要素，推动城市公共空间真正走向人们的日常生活，使城市成为人民的城市。

Two designers, Liu Enfang and Fan Wenli, regard "good urban design" as an important tool and essential element to support and express the spirit of the city (non-material goals) at the material space level. Based on the consistent design philosophy of "openness, interconnection, sustainability, mobility and human nature", the daily power of urban public life is shaped from the heart.

Adhering to the "Urban+" breakthrough thinking and method, design strategy and implementation control, the team of Liu Enfang and Fan Wenli completed series of acclaimed design creations, which include the Xiong'an architectural style guide, the urban design of the Shanghai Hongqiao High-speed Railway Hub Business Core District (Phase I), and the central enterprise in the B area of the Shanghai World Expo. Headquarters base urban design, etc.

We have being paying attention to "the power of daily life" over the years, as it is a crucial dimension of the enhancing quality of urban public open spaces. Through urban design, urban public domain can closely integrate with urban life, which allows the high-quality open spaces to be the welfare of more citizens. To combine this waterfront theme, the "eleven urban-design strategies" construct the positive and healthily connection of waterfront spaces and people. It accelerates the waterfront open spaces to be a daily part of urban life and move the city towards a city for people.

问题
公共空间为谁设计？

策略　具有包容性

西岸港口滨江规划暨概念竞赛城市设计项目

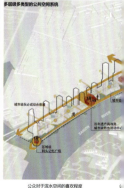

多层级多类型的公共空间系统

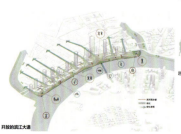

开放的滨江大道

公众对于滨水空间的喜欢程度

城市滨水公共空间——如何成为适合所有人的地方
　　城市滨水公共空间应致力于成为满足人们各种需求的有吸引力的地方。光有物理要空间是不够的，还需要空间具有内涵，有人本关怀以及最广泛的使用性。
　　城市滨水公共空间的设计如同真正关注全龄段、面向所有人！我们以为在空间设计方面，首先需要做的就是将公共空间进行分级别、分维度的，城市级、社区级、区级级、干预规划的公共空间和城市所服的对象各各有所侧重，空间尺度不同，将承载的社会活动的类型、能次也不同，以社区级公共空间为例，应主要关注最直行，满足家庭的生活品质和生活需要。
　　我们看到了解城市中的各类人群的需求，继续发展更广的用途和创造性的生活方式，使其不仅是一个有吸引力的工作场所，还是一个充满活力的24*7的生活场所。

Urban Public Domain - A place for everyone. Our urban public space is dedicated to becoming an attractive place to meet people's various needs. In recent years, Shanghai has implemented multiple strategic plans, such as riverside corridor. It means the urban waterfront public space will have the opportunity to become a more functional and dynamic space.
The design of public space focuses on all age groups and all people. In the perspective of space design, the first thing that needs to be looked at is that public space should be sorted hierarchically including city level. District level, community level. Public space of different hierarchy will confront different

越南人口构成年轻化　　越南就业人口

P302

《空中造楼机》课题组

Building Machine in High Space Research Group

回归水·泥，未来建造

Return to Water & Cement, Construction in Future

石头粉碎煅烧为水泥是一种蝶变，水泥遇见水凝固是一种回归，这是人们近两百年来善用水和石头等自然资源的建造智慧。

绿色建筑的核心是节能、低碳和工业化，建筑产品的标准是安全、健康与可持续。"回归水·泥，未来建造"装置试图回答钢筋混凝土建筑工业化的途径不只是预制装配，基于机械化施工、自动化控制、信息化管理的现浇钢筋混凝土建造技术也是一条重要的路径，尤其是面对大中城市中量大面广的高层和超高层居住建筑。

装置创意源自"十三五"国家重点研发计划"绿色建筑与建筑工业化"，"建筑工程现场工业化建造集成平台与装备关键技术开发项目"之课题"超高层建筑落地式钢平台与设备设施一体的智能化大型造楼集成组装式平台系统"(简称"空中造楼机")的初步研究成果，为真实装备的1:10模型。

技术核心为自动开合的装配式模架模板系统、自动升降的建造平台系统和混凝土浇筑养护系统。

Stone crushing and calcining into cement is a kind of rebirth. Consolidation of Cement encountering water is a kind of return. This is the building wisdom that people have used natural resources such as water and stone for nearly two hundred years.

The core of green building is energy conservation, low carbon emissions and industrialization, and the standards for whole building product is safety, health and sustainability. The "Return to Water & Cement, Construction in Future" installation attempts to answer the path of industrialization to reinforced concrete buildings. It is not just prefabricated,the cast-in-place reinforced concrete construction technology based on mechanized construction, automation control and information management is also an important path. Especially, the large or medium-sized city has a large number of high-rise and super high-rise residential buildings.

The installation's concept originated from the initial results of project of "13th Five-Year" national key R&D plan. This is the 1:10 model of real equipment.

The cores of the technology include the construction platform system with automatic rising or falling, the concrete formwork system with automatic opening and closing, and the concrete pouring & maintenance system.

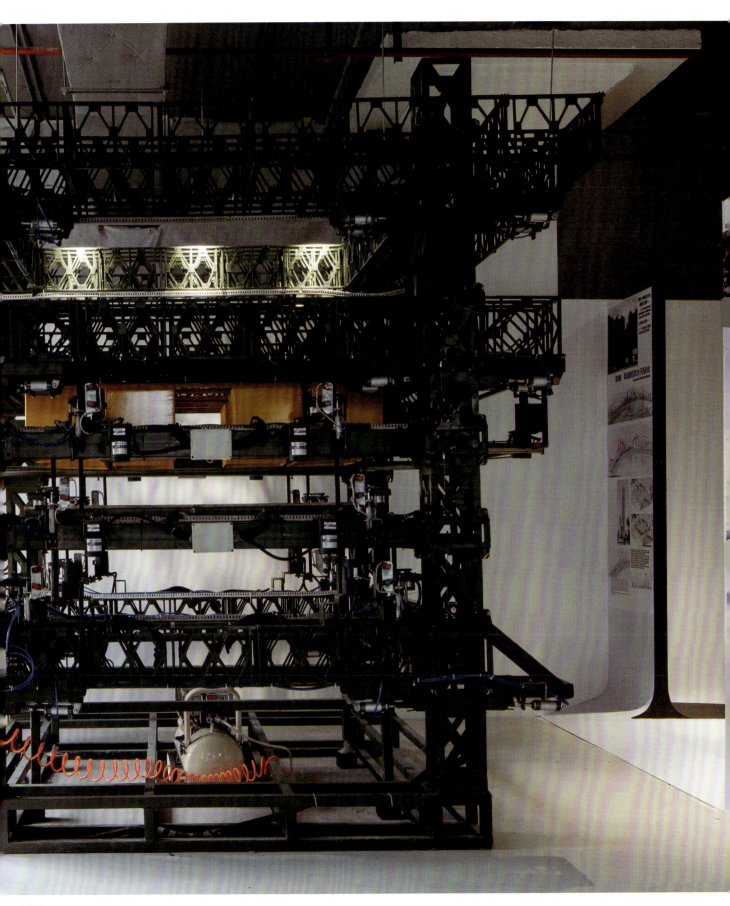

P303

遇水而柔，逢道则刚
Softness from Water, Toughness from Dao

匡晓明
Kuang Xiaoming

人与城市和自然共同构成了有机生命体，城市建设就是人类智慧和自然环境共同作用的结果。而人工化与自然化的理性整合正是建筑设计的核心议题。唐代画家张璪所言"外师造化，中得心源"给我们以启示。"师法自然，意行人工"可以成为一种营城之道。以此为鉴，"遇水而柔，逢道则刚"成为我们上海崇明智慧岛孵化器的设计理念，遇水逢路则柔刚并运，表现出建筑设计在应对不同外围自然与人工环境时所采取的外刚内柔的空间因应策略。

Man, city and nature constitute organic life body together. Urban construction is the result of human wisdom and natural environment. The rational integration of artificial and natural is the core issue of architectural design. Zhang Yu, the painter of Tang Dynasty, said that "learning from the outside world, learning from the heart" gives us inspiration. "Learning from nature and doing things with will" can be a way to build a city. Taking this as a example, "when meeting water, it is soft, when meeting road, it is just" has become the design concept of our incubator of Chongming intelligent island in Shanghai. When meeting water, it is soft, when meeting road, it is hard and when it is transported together, it shows that the architectural design adopts the outer rigid and inner flexible space response strategy in response to the difference between the outer natural and artificial environment.

P304

唤醒一座城市的森林
Forest Awakens the City

沈阳建筑大学HA+ STUDIO
Shenyang Jianzhu University HA+ STUDIO

艺术装置以两个固定的三角形，分别代表自然和人类活动。通过两个三角形不同角度的观察结果，探讨人类活动对于河流生态景观的影响。究竟什么比例的人类活动是适宜的？什么尺度的生态种植可以使得城市公园真正成为城市绿核？如何营建具有自我演替能力和韧性的智慧生态？我们希望通过介绍沈阳浑河一河两岸生态化种植的具体策略，通过生态化种植方法完成公园景观带向生态式景观带转变，通过城市森林去唤醒一座城市的活力。

Two fixed triangles are used to represent the natural and human activities. Based on the observation results of two triangles from different angles, this paper discusses the impact of human activities on river ecological landscape. What proportion of human activities is appropriate? What scale of ecological planting can make urban parks become urban green cores? How to build a smart ecosystem with self succession ability and resilience? We hope to introduce the specific strategies of ecological planting on both sides of Hunhe River in Shenyang, complete the transformation from park landscape belt to ecological landscape belt through ecological planting method, and wake up the vitality of a city through urban forest.

■ 概念生成 Concept generation

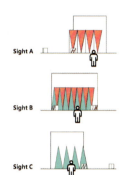

通过两个三角形不同角度的观察结果，探讨人类活动对于河流生态景观的影响。

Based on the observation results of two triangles from different angles, the influence of human activities on river ecological landscape was discussed.

Sight A

Sight B

Sight C

SIGHT

A

B

C

■ 概念生成 Concept generation

BEFORE

浑河两岸生态状况总体处于低干扰状态，生态效益较稳定。

The ecological situation on both sides of Hunhe River is generally in a low disturbance state, and the ecological benefits are relat-ively stable.

NOW

随着城市发展，浑河生态种植出现破坏-重构-稳定的生态群落更替过程，人类活动处于高干扰状态。

With the development of cities, the ecological planting of Hunhe River has been destroyed, reconstructed and stabilized, and human activities are in a state of high disturbance.

REFLECTION

人类应思考，市民活动与城市内河生态公园之间相互影响的辩证关系。

Human beings should think about the dialectical relationship between civic activities and urban inland river ecological parks.

P305

同济大学建筑系 2019 本科
毕设——陆家嘴世纪大道实验
2019 Graduation Design of Tongji Univ.
CAUP Undergraduates — Experiment on
Lujiazui Century Avenue

同济大学建筑系 2019 本科毕设——陆家嘴世纪大
道实验课题组：蔡永洁、许凯、贾涛
Team of 2019 Graduation Design of Tongji Univ. CAUP
Undergraduates — Experiment on Lujiazui Century
Avenue: Cai Yongjie, Xu Kai, Jia Tao

中国的很多新区在上一个阶段城市发展中，对适宜的尺度、混合的
功能、城市活力等要素考虑甚少。课题组将带领学生对"世纪大道"
这一上海的标志性"问题道路"进行实验性改造设计及畅想。根据
种种问题分析和国际案例比对，课题组为世纪大道设计了两种改造
畅想方案，一种将世纪大道下沉；另一种将世纪大道抬升。展览时
在呈现详细的设计图纸和分析图及效果图的同时，又将最终的设计
通过 1:1000 的大比例模型形式加以呈现。

In many cities in China, in the development of the previous stage,
there are few considerations for appropriate scales, mixed functions,
and urban vitality. The teaching team will lead students to carry out
experimental design and imagination of Shanghai's iconic "problem
avenue". According to various problem analysis and international case
comparison, the research team designed two kinds of transformational
ideas for Century Avenue, one to sink the Century Avenue; the other to
lift the Century Avenue. At the same time of presenting detailed design
drawings, analysis charts and renderings, the designs were presented
in a large scale model of 1:1000.

P306

剪水
Conquering the Land from Water

南京大学丁沃沃
UALab 团队
Urban & Architectural
design & research Lab

千年奔腾不息的河流连通了全世界的水域，滨水建筑以倒影的形式将自身的变迁书写在水面上，以水为纽带连接了全世界无数的滨水建筑，使它们在水中相遇相知。无论是柔美秀丽的古典建筑，还是刚硬挺拔的现代建筑，水中倒影的灵动使古典与现代浑然一体，不再对峙。水面似静，涟漪自动，静动交织，自成佳趣。本次展览通过设计一个视错觉装置，展现各种各样的城市建筑在水边的倒影，用静态的模型呈现出动态的视觉感受，表达水的力量，滨水建筑的魅力。

The millennium-prosperous river connects the world's waters, and the waterfront buildings write their own changes on the surface of water in the form of reflections. Water, as a bond, connects countless waterfront buildings around the world, allowing them to acquaint with each other in water's embrace. Whether it is a gently beautiful classical building, or a rigid and straight modern building, the reflection in the water can emerge the classical and the modern as a whole and leave the confrontation between them behind. The water seems to be static, while subtle ripples spread out spontaneously, and this static-dynamic status brings a unique interest. Through the design of an optical illusion device, this exhibition presents the reflection of various urban buildings at the water's edge. A dynamic visual experience is provided with the static model, expressing the power of water and the charm of waterfront buildings.

镜面有机玻璃

有机玻璃灯箱

城市画像

镜面木板

黑色玻璃纤维塑料框

机械合金转盘底座

镜面有机玻璃

木质烤漆展台

剪水 Conquering the Land from Water

P307

王云、汤晓敏团队
Wang Yun & Tang Xiaomin

记忆·融合·共生——城市滨水景观更新规划与设计

Memory & Fusion & Symbiosis — Renewal Planning and Design of Urban Waterfront Landscape

因水而兴的长三角地区，正面临着水环境恶化、滨水生态与活力空间缺失等问题。镇江古运河、上海黄浦江、昆山夏驾河等项目探索了大型滨水空间的景观化、生态化与人文化的路径与策略；昆山致塘河、丹阳九曲河、嘉兴嘉善塘等小型滨水公共空间的再生与精细化设计，提出了城市高密度区域"金边银角"的公共绿地更新计划；昆山中心城区海绵城市与蓝绿统筹体系规划探索实现了水利安全与生态、水乡文化与记忆、海绵城市理念与适用技术的融合与创新。

The Lower Yangtze Delta is a region with abundant water network; water resource is particularly significant to the prosperity of the region. However, water environment deterioration worsened in recent decades, such as lacking ecological and attractive waterfront spaces. Zhenjiang ancient Cannel, Shanghai Huangpu River and Kunshan Xiajia River projects explored large-scale riverside landscape improvement using integrated landscaping, ecological and cultural approaches and strategies. Small-scale waterfront public space renewal and delicate design, such as Kunshan Zhitang River, Danyang Jiuqu River and Jiaxing Jiashan River projects, proposed the "Golden Edge and Silver Corner" public greenspace regeneration plan for the high-dense urban area. The planning of greenspace and water network in central Kunshan city established integrated approaches to realize multiple goals, including enhancing water conservancy safety, local cultural identity and spatial memory, landscape resiliency and technique innovations.

太仓港区七浦塘生态公园规划设计
● 上海市优秀工程设计二等奖
公园位于以"水城绿链、市民共享"和"飞阁流丹、覆飞叠映"为战略与景观定位，构筑多样生境，强连系、添、湿、草、花、林交融的活数因弦队，内部融合、外部联通，构筑外延内倾的康居空间格局。

武汉未来科技城龙山溪气候适应性与韧性设计
● IFLA-APR文化与城市景观奖-荣誉奖
设计基础适应气候变化与场地设计的理念，探讨生态对特殊性、湿地自然与城市间的联系、场地可持续性、清和人与场地的关系、生活可持续性、追踪石感与人之间的联系，在2016年武汉50年一遇洪涝灾大洪水期间，适地成功缓解了泊洪的压力。

康体健身的活力绿带：上海松江油墩港沿河休闲绿地设计
● 上海市优秀勘察设计二等奖
油墩港是上海市规划的"一环十射"骨干级航道建设中的重要航道，该段河岸由及公共绿地状态修理器，以"南北贯穿、西西东连"的空间网络，形成了一条景观生态型的中央绿色健身与休闲功能的活力绿带，丛构建了生态毗邻松江区域绿道，它打造了四季缤纷的休闲及后应自然生态性，儿童友好的亲子共享空间，公采用信息共管理措为，构建了快的游务的社区康养示范园。

P308

江南水
Water of the South of the Yangtze Region

思作设计工作室（范文兵、张帆）
Atelier Fan (Fan Wenbing, Zhang Fan)

江南水用抽象与具象相结合的手段，让观众通过体验和视觉，感知作者对"江南之水"的印象。

Jiangnan water uses the combination of abstract and concrete means to let the audience feel the author's impression of "Jiangnan Water" through experience and vision.

张永和／非常建筑
Yung Ho Chang / Atelier FCJZ

水之居学——中国美术学院良渚校区
Live-Learn on the Water — Liangzhu Campus of China Academy of Art, Hangzhou

良渚新校区将成立创新设计学院，迁入艺术管理与教育学院、继续教育学院，形成"三学院"格局。全日制在校生3000人，继续教育学生1000人。良渚校区的建设将突出四个"面向"。面向未来：强调面向未来的跨学科、复合型的创新人才培养，推动设计与信息经济、人工智能的深度融合，打造当代设计教育最前线。

良渚校区是面向未来的创新校园，同时又需回应强调手工劳作与实践操作的艺术类大学的教学模式；通过对教学体系的研究促进对传统教学空间进行创新改革，将功能固定的"教室"发展为空间多义的"工坊"。绵延的工坊为未来教学的发展与融合提供多种可能。

宿舍－兴趣社－工坊的垂直式分层体系，将生活与教育空间相互融合，塑造出以兴趣为纽带的社区化的空间，体现"学院即社区"的概念。"居学"的概念是指导良渚校园整体布局的核心思想，是对良渚校区如何成为大数据、智媒体时代的一所面向未来创新校园的积极回应，是"生活即教育，学院即社区"的理念的具体设计表达。

Liangzhu Campus is the new campus of China Academy of Arts, new home for College of Innovative Design, College of Art Management and Education, and College of Continuing Education. It will house 3000 full-time students and 1000 continuing education students. The four main objectives of this new campus include development of interdisciplinary innovative talents; promote design and informational economy; integration of artificial intelligence technology; and to set bench mark for contemporary design education.

Design of the Liangzhu Campus aim to create innovative teaching spaces for the future, and to promote manual work and practical operations at the same time. We do so by redefine the traditional teaching spaces and transform the typical "classroom" into "workshops" with multiple functions. Prolonged workshops provide possibilities for diversified teaching modes.

The vertical stratified system of dormitory, social space and workshops creates a community that embodies the concept of "college as community". This is the core idea guiding the overall layout of Liangzhu Campus. It is a positive response to how Liangzhu Campus becomes a future-oriented Innovative Campus.

P310

法国 TER（岱禾）景观
与城市规划设计事务所
Agence TER (France)

鳗鲡嘴儿童乐园——保留
多样共享的城市公共空间

Children's Hill —
Manlizui Childs' Playground

鳗鲡嘴滨江绿地是黄浦江东岸滨江公共空间的其中一段，拥有两处高地与一处大型开阔草地，在这里不仅可以望见浦西美景，也可观赏浦东风光。

未来的城市空间，丰富的绿化种植空间需要与城市其他空间良好互动，为城市提供活动及活力的多样共享空间。鳗鲡嘴儿童游乐空间的设计不拘泥于公园本身，公园的轮廓最终融入上海这座城市的天际线中。

整个公园即是扩大的儿童游乐空间：宽阔的草坪——孩子们在最自然的游乐园中嬉戏玩耍，大型户外滑梯——是整个游乐场的点睛之笔。儿童游乐空间在这里与城市公共空间浑然一体，而儿童游玩的场景自然而然地成为了公共空间的风景，我们期待小朋友们在这里创造自己的小世界。鲜亮的色彩再次激活浦江东岸。

Manlizui riverside is one of the urban spaces developed by Agence TER on the eastern bank of the 22km Huangpu river park masterplan and development. It is sculpted by an artificial topography of two-coloured highlands and vast lawns. As a miniature metaphor of the river overlooking Puxi bank and enhancing the new Pudong skyline. Attentive profiling of space offers privileged views of the river and the metropolitan landscape. Natural spaces and ecosystems have been created to welcome and protect the river's fauna and flora biodiversity.

This design is highly contrasted; the sculptural coloured geometric volumes stand out in a luxurious green setting. This interaction of the natural areas and the hard surfaces provides the city with a variety of spaces, activities and intensities.

The park is more than a playground: the wide lawn is an accessible area for all generations. The playground is an essential compartment of the urban public space, and it has an important see and to be seen characteristic. The lively playing that takes place here is a unique scenery on the riverside with its buzziness and sculptural slide. According to the designer's expectations, children are creating every day a whole new imaginary and colourful world here.

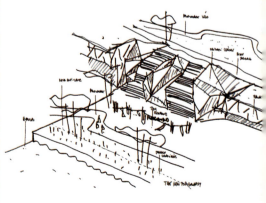

P401

Let's Play, 天真无邪儿童展
Let's Play, Innocence Kindness Children's Exhibition
Let's Play, Innocence Kindness Children's Exhibition

策展人：戴春
学术支持、系列论坛、活动策划团队：Let's Talk 学术论坛（金怡、董艳、董林涛、彭柏寒）
策展运营、展陈设计及布展团队：
筑术空间（虞晓宸、王佳佳、吴薇、谭雅秋）

Curator: Dai Chun
Academic Support, a Series of Forums, Event Planning Team: Let's Talk Academic Forum (Crystal Jin, Dong Yan, Dong Lintao, Peng Baihan)
Exhibition Operation, Exhibition Design and Exhibition Arrangement Team: Archiepos Studio (Ignatius Yu, Wang Jiajia, Wu Wei, Tan Yaqiu)

城市不仅是成年人的世界，今天的孩子未来会成为城市的主人。我们希望在展览中让孩子们了解今天的建设者为滨江贯通付出的努力，人们如何把曾经苍白、污染、混乱的空间变成宜人的城市景观。我们更希望传递出"如果这个不够美好，就让我们一起改变它"的理念，并让孩子们有机会在艺术季中展现惊人的创造力。

一本巨大的沉浸式童书
整个展览将布置成一个由充满滨江空间特色元素组成的简笔画背景。随着展览的推进，我们将组织孩子们对该空间填色创作。完成自己心目中最美滨江的创作。

专属活动时间
儿童展区将定期开展系列艺术教育活动，如小小导览员、上海本土滨江动植物认知体验课、空间研究工作坊等寓教于乐的亲子课程，加深对城市空间艺术季及滨江空间改造的认知。

Cities are not only the world of adults. Today's children will become the masters of the city in the future. In the exhibition, we hope that children can understand the efforts made by today's builders for the riverside connection and how people have turned the once boring, polluted and chaotic space into a pleasant urban landscape. We want to convey the idea of "If this is not good enough, let us change it together" and give children the opportunity to show their amazing creativity during the art season.

A huge immersive children's book

The entire exhibition will be arranged as a background of stick figures filled with the characteristic elements of the riverside space. As the exhibition progresses, we will organize children to colour the space, completing the most beautiful riverside creation in their mind.

Exclusive event time

Children's exhibition area will regularly carry out a series of art education activities, such as children's guides, Shanghai native riverside plant and animal cognitive experience class, space research workshops and other parent-child courses to deepen cognition to deepen the understanding of urban space art season and riverside space transformation.

P501

缘生缘灭——江西路『自来水桥』
Demolish & Reborn —— Kiangse Road Water Bridge

缘灭

1883 年，英商上海自来水公司兴建的杨树浦自来水厂实现供水。为将自来水送至苏州河南岸的英属租界区，借助具有"教堂街"之称的主要南北向道路江西路修建了跨苏州河水管并伴行建桥，俗称"自来水桥"，兼有车辆通行功能。同时，在江西路、香港路口建造了一座 31.5 米高的配套水塔，以提升水压，实现远程输水。

据档案记载，自来水桥"在抗战期间被敌伪拆除，行人车辆均绕道四川路桥或河南路桥而行，致两桥交通异常拥挤"。而后时局变迁，江西路自来水桥再未重现，终成一段记忆残影。

缘生

2019 年，在苏州河两岸贯通工程的研究过程中，从提升苏州河河口区域行人步行感知质量的角度出发，设计团队对以步行桥方式重建"自来水桥"的可能性进行了研究，并对其与周边历史街区的重构关系进行了想象。未来的江西路步行桥不仅局限于通行这一单一目的，还承载了休闲、城市雕塑等功能，在外形上既对消失的"自来水桥"有所呼应，又体现了时代技术的进步。

城市基础设施在历经城市变迁的过程中往往因发展机缘的丧失而消逝，而这一段曾经存在的记忆影像反过来又为城市基础设施在未来的机缘重生留下了无法忽视的在地线索。城市在空间物的相遇与分离间孕育出关于未来的美好憧憬……

以空间装置的方式展现苏州河边的"自来水桥"从过去到未来的虚拟化存在，艺术性地记述它与河、与人的机缘关系！

Kiangse Road Bridge Demolished

In 1883, the Yangtszepoo Waterworks built by Shanghai Water Works Co., Ltd., started water supply. In order to send tap water to the British concession area, which is on the south bank of the Suzhou Creek, a "tap water bridge" with a cross-river water pipe was built along with the main north-south road Jiangxi Road, which is known as the "Church Street". It also has the function of vehicle access. Meanwhile, a 31.5-meter tall water tower was built at the intersection of Jiangxi Road and Hong Kong Road to increase water pressure and achieve remote water delivery.

According to the records, the water bridge was "destroyed by the enemy during the period of Anti-Japanese War. Pedestrians and vehicles are bypassing the Sichuan Road Bridge or Henan Road Bridge, resulting in traffic congestion between the two Bridges." After the change of the situation, the Jiangxi Road tap water bridge has not been rebuilt, and eventually became a memory.

Jiangxi Road Bridge Will be Reborn

In 2019, during the research project of connecting the two sides of Suzhou Creek, the design team studied the possibility of rebuilding the "tap water bridge" in the form of pedestrian bridge from the perspective of improving pedestrian perception quality in Suzhou Creek estuary area, and imagined its reconstruction relationship with the surrounding historical blocks. The future Jiangxi Road pedestrian bridge is not only limited to the single purpose of traffic, but also carries functions such as leisure and urban sculpture, which not only echoes the disappearing "water bridge" in appearance, but also reflects the technological progress of the times.

In the process of urban transformation, urban infrastructure often disappears due to the loss of development opportunities, and this memory image that once existed in turn leaves local clues that cannot be ignored for the rebirth of urban infrastructure in the future. Between the meeting and separation of space objects, the city breeds a beautiful vision of the future.

Show the virtual existence of this bridge from the past to the future in the way of space installation, and describe its relationship with the river and people artistically.

杨明
案例研究团队：宿宸、袁野、刘济维、冯钰、宿新宝、苏萍、王天宇
展示模型制作：上海波海建筑模型设计制作有限公司

Yang Ming
Research Team: Su Chen, Yuan Ye, Liu Jiwei, Feng Yu, Su Xinbao, Su Ping, Wang Tianyu
Modeling: Shanghai Bohai Architectural Model Design and Production Co., Ltd.

P502

切片苏州河
Slices of the Suzhou Creek

同济原作设计工作室
（张姿、章明、王绪男、丁阔、丁纯、张林琦、刘炳瑞、林佳一、王祥、岳阳、鞠曦、刘皓、李泊衡、吴屹豪）
Tongji Original Design studio
(Zhang Zi, Zhang Ming, Wang Xunan, Ding Kuo, Ding Chun, Zhang Linqi, Liu Bingrui, Lin Jiayi, Wang Xiang, Yue Yang, Ju Xi, Liu Hao, Li Boheng, Wu Yihao)

同济原作设计工作室主持参与了苏州河南岸黄浦区段的公共空间提升改造以及杨浦滨江贯通工程项目。本次展览依托苏州河项目实践的经验与感悟，针对水岸环境有机更新的功能和空间状态，以切片的方式展现水岸再生与城市生活、空间的关系。

展览采用极简的手法来抽象空间与图像，聚焦"水与岸相遇"的空间主题；在展墙墙面翻出一条"窗口"，露出蓝色的、黏贴有苏州河南岸黄浦段总平面图的凹龛，并选取了若干个重要节点设置可抽拉的河岸断面切片，断面切片以透明亚克力丝网印刷线描图纸的方式展示。

观展者在"沿河行走"中可自行抽出这些断面，除了看到水岸与城市空间的水平耦合关系外，平时不为人熟知的市政管网、防汛墙、泵站等基础设施与城市公共空间在垂直空间的复合再生状态也展现在观展者面前，透明的展示介质也让一种叠合的观察状态成为可能，在纯粹的展示背景中凝视滨河城市更新的愿景将带来视觉冲击并引发对滨水空间的思考。

Tongji original Design studio has hosted the waterfront renewal of the Suzhou Creek and the Yangpu Waterfront connection project. Based on the experience and sentiment of the Suzhou Creek project practice, this exhibition, in response to the issues of functional and special renewal of waterfront environment, aims to show the relationship between waterfront regeneration and urban life.

The exhibition uses a minimalist approach to abstract space and images, focusing on the theme of water- meet-the-shore. The core of the exhibition is a blue light strip that extends over the wall and penetrates the entire surface of the channel. The General Plan of the South bank of Suzhou Creek in Huangpu district is lay on the background of the light belt. Several important nodes, drawings of which are printed on acrylic panels, are selected to set up the river bank sections.

During the process of "walking along the river", the visitors can pull out these panels and see the composite growth of the space in the horizontal and vertical direction after the waterfront regeneration; infrastructure, including the municipal pipe network, the tamper-proof wall, the warehousing, etc., will be showed in the sections, which is usually unfamiliar to the visitors. A new superposition way of seeing in front of a purified background will lead to a visual impact and thinking of the waterfront space.

P503

牟振宇
Mou Zhenyu

消逝的城市之河
Vanishing Urban Watercourses

开埠以后，上海开始了城市化进程，原有河浜体系受到巨大冲击。上海的城市化进程可描绘为由"五里七里一纵浦，七里十里以横塘"的河浜圩田景观向"纵横界画似棋盘"的街道洋房景观转变的过程。在1900年自1930年间，租界的辟建、扩张以及越界筑路使传统水乡的河浜体系遭到破坏，大部分河浜被截流筑路或填埋筑路。本展览在爬梳河道变迁的基础上，通过深入挖掘河浜变迁与城市化的内在联系，以动画的方式，透析近代上海城市化演进的机理。

Shanghai has experienced urbanization since it was named a treaty port, which changes the original watercourse network greatly. The urbanization in Shanghai could be described as a transformation from the rural landscape of interlaced watercourse network to the urban image of chessboard morphology. Between 1900 and 1930, many watercourses were replaced by the roads for the extension and construction of the Concessions. This exhibit investigates the nexus between the urbanization and the watercourse changes to illustrate and analyze the nature of the urbanization of Shanghai by animation.

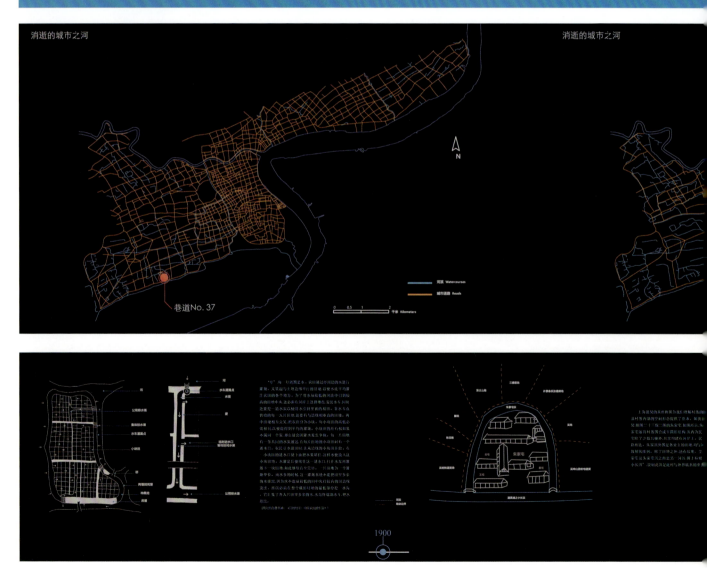

消逝的城市之河

消逝的城市之河

巷道No. 37

河浜 Watercourses

城市道路 Roads

1900

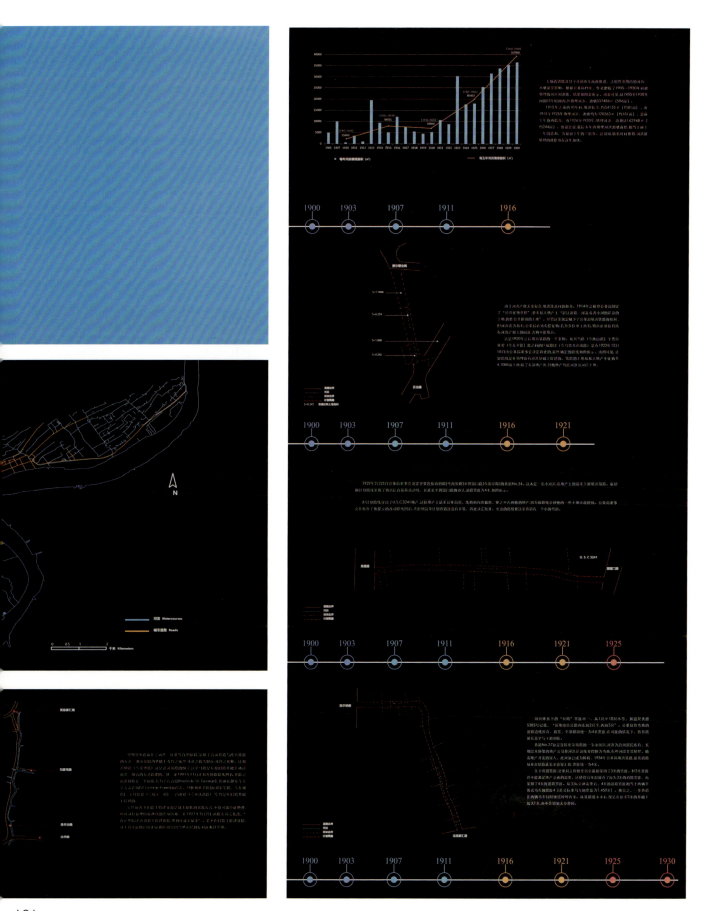

P504

薛鸣华、王林团队
AMJ & SJTU CITY LAB

虹口·北外滩
北苏州路滨河空间
城市设计——VR
浸入式设计效果体验

The North Bund North Suzhou Rd. Riverside Urban Design — Design Effect with VR Device

在"一江一河"贯通的背景下，以城市更新的方法，对北苏州路沿线滨水空间进行提升。设计以成为"最美上海滩河畔会客厅"为目标，将北苏州路滨河空间断点贯通，并将空间划分为上海大厦活力花园段、宝丽嘉酒店休憩观景段、邮政大楼风貌展示段与河滨大楼特色风情段，形成"一岸四段"的总体规划结构，并将原有的"人行道＋车行道＋滨河栈道与绿化"的路面组织形式，提升为"步行街＋共享街道＋观景平台＋滨河步廊"的新模式。在满足防汛墙结构不变、特殊车辆可通行的基本条件下，设计上做到全面无障碍贯通滨河空间、大幅度增加人行空间、多模式消解高差、多节点特色打造、多角度视觉引导与相应的精细化管理导则，整体提升滨河空间活力。

Under the background of the development plan for Huangpu River and Suzhou Creek, the riverside space along the North Suzhou Rd. will be upgraded by the method of urban renewal. The design aims to become the "most beautiful riverside living room" as the goal, and the breakpoint of the riverside space of North Suzhou Road is connected. The space is divided into the Broadway Mansions vigor garden section, the Bellagio Hotel leisure landscape section, the Shanghai Postal Museum historical section and the Embankment Building characteristic section. Forming the master plan structure of these four major themes. The road organization form of "Pavement + Roadway + Riverside Road and Planting" was upgraded to a new mode of "Pedestrian Street + Shared Street + Viewing Platform + Riverside Corridor" to enhance the vitality of the riverside space. The design seeks to provide free access to a riverfront that elevates the pedestrian status of the shoreline to a circulation node that provides sceneries and visual guidance trough different heights along the course. This characteristic also helps to satisfy the safety requirements of the flood prevention wall and the passage for special vehicles. Design in conjunction with refined management guidelines will enhance the vitality of the riverfront.

182

P505

乔治·乔尼庚
团队（意大利）
Team of Giorgio
Gianighian (Italy)

展览由三个不同的叙述组成，分别被称为"历时性""共时性"和"视觉"，每一个都展示了威尼斯这座城市的不同方面。威尼斯的主要使命是与它所处的水域保持不可分割的关系。

历时性路线通过互动地图展示了城市的发展，突出了主要的建筑成就：教堂、公共建筑、宫殿和房屋。这里有五幅地图，每一幅都描绘了威尼斯生活的一个特别重要时期，直至当代。

"共时性"的同步路线显示了一些主要问题及其解决方案，威尼斯不得不面对能够生存下去的人居环境，泻湖的控制由连接超过几百个小岛共同组成，建筑技术和淡水供给共同发挥作用。

所有这些都通过幻灯片、面板和交互式地图进行了说明。

最后一部分，还配有幻灯片、印刷面板和视频，展示了这座城市的各种元素和生活片段，比如：无数海滨中最重要的：建造和划一艘贡多拉！这是这座城市在世界各地最受欢迎的特色之一，可以欣赏到沿着大运河的宫殿和其他景观。

The exhibition consists of three distinct narratives, called "diachronic", "synchronic" and "visions", each showing different aspects of a city, Venice, whose main vocation is a indissoluble relationship with the waters from where it rises.

The diachronic route shows the development of the city in time, through interactive maps where the main architecture accomplishments are underlined: churches, public buildings, palaces and houses. There are five maps, each for a specially important period of the life of Venice, up to contemporary days.

The synchronic route shows some of the main problems (and their solutions) that Venice had to confront to be able to survive, from the control of the lagoon to the connection between the more than hundred islets making up the city, including analysis of building techniques, collection of fresh water et alia.

All this is illustrated with slideshows, panels and interactive maps.

The last section, also illustrated with slideshows, printed panels and with videos, shows various elements and episodes of life in the city, such as: the most important of the countless waterfronts; building and rowing a gondola, one of the most popular feature of the city around the world, views of the palaces alongside the Grand Canal, and others.

P601

浪潮
Overwhelming

上海大界机器人（梁喆、秦川、纪美琳、玛丽亚·拉日维纳、赵迪）

RoboticPlus.AI (Liang Zhe, Qin Chuan, Ji Meilin, Maria Razzhivina, Zhao Di)

浪潮不可倒流，正如时光不可逆转。——奥维德

我们身处的此时此地持续涌动着无数新奇、随机、变幻莫测、势不可挡的相遇。纯净的自然生态遇见人类活动，传统的建筑材料遇见新兴的建造技术，不断深度学习中的人工智能发展遇见不可名状的人类感官体验；而当科技浪潮涌向建筑行业，大界机器人应运而生。

人类在北极冰芯中发现了塑料颗粒，人造材料被错置于自然的深处，我们似乎再也无法觅得人类未曾染指的自然。怀揣着对生存现状的敬畏之心，之于材料、之于科技，我们在抗拒中接受，在迷惑中探索。直面不可逆转的现代文明和人类发展需求，艺术便是我们回应思考的一种途径。

本装置呈现的正是这种思想的延续。"overwhelming"本意为淹没感，势不可挡，如海浪一般气势磅礴。借助于机器人建造技术，波浪状的装置连绵起伏、极具流动感，结合互动性的光影效果，营造出具有沉浸感的似水空间，供我们思考这一切因相遇而发生的碰撞、融合、启示。

Neither can the wave that has passed by be recalled, nor the hour which has passed return again. — Ovid

The moment which we currently share is constantly filled with numerous novel, random, dynamic and overwhelming encounters. Pure and natural ecological environment encounters with human activities, old construction materials encounter new building techniques, deep-learning artificial intelligence encounters with unprogrammable human sensation. When the wave of technology heads towards construction industry, RoboticPlus.AI came into being.

Human beings have found microplastic particles in ice cores drilled in the Arctic, artificial material was misplaced into the deep nature, we can no longer discover authentic nature which is untouched by human. With all respect to the current survival environment, speaking of materials and technology, we accept with reluctance, explore with confusions. Facing irreversible modern civilization and human development demands, art is chosen to be a pathway for us to respond.

This installation on display is an extension of such thought. The word "overwhelming" originally refers to the sense of flooding, the momentum that cannot be stopped, which is comparable to the magnificent sea wave. Benefitted from robotic fabrication techniques, this wavy installation has a rolling and highly fluid appearance. The immersive watery space is created with the integration of interactive lighting effects, giving us opportunities to think about all these occurrences of confrontation, interaction and enlightenment arising from encounters.

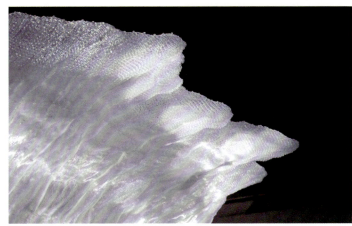

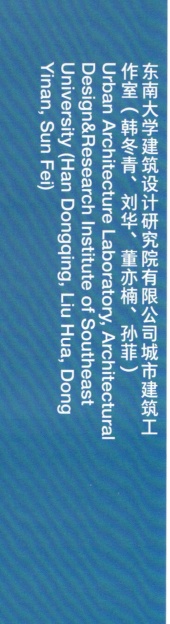

P701

十里秦淮·西五华里——
南京内秦淮河
西五华里滨河地段城市设计

东南大学建筑设计研究院有限公司城市建筑工
作室（韩冬青、刘华、董亦楠、孙菲）
Urban Architecture Laboratory, Architectural
Design&Research Institute of Southeast
University (Han Dongqing, Liu Hua, Dong
Yinan, Sun Fei)

Inner Qinhuai River & West Half Ten
Miles — Urban Design on Riverside
Area along West Half Ten Miles of
Qinhuai River, Nanjing

城市设计编制起讫时间：2016.6.3—2017.7.30
城市设计通过专家论证时间：2018.11.14
南京市规划和自然资源局时间：2019.1.30
规划面积：22.7 公顷

"十里秦淮"是南京的母亲河，与中华路构成南京城市起源发展的
关键轴线，其沿河地段将成为南京市全域旅游最重要的载体之一，
是老城南整体复兴的重要纽带。西五华里指"十里秦淮"的西半段。
现状河道狭窄，两岸用地权属复杂，公共认知度偏低，临河开发建
设范围局促。

城市设计依托既有历史遗存与历史信息，从老城南整体保护展示的
角度，串联沿线历史文化资源及生活服务设施，积极塑造南京重要
的历史文化线路，提升旅游配套设施品质，同时完善民生服务设施。

Starting and ending time of urban design compilation: 2016.6.3 — 2017.7.30
Date of urban design passing expert argumentation: 2018.11.14
Date of Nanjing Bureau of Planning and Natural Resources: 2019.1.30
Planning Area: 22.7ha

"Ten-miles of Inner Qinhuai River" is the mother river of Nanjing, constitute the key axis of Nanjing's urban origin and development with Zhonghua road. Its riverside area will become one of the most important carriers of the whole region tourism in Nanjing. It is an important link for the overall revival of the south historical districts in Nanjing. "West half ten miles" refers to the western half of the ten-miles inner Qinhuai River. The current channel is narrow, the ownership of lands on both sides is complex. Public awareness is low, and development and construction scope of riverside area is limited.

Urban design relies on existing historical relics and historical information. From the perspective of the overall protection and exhibition of the south historical districts of Nanjing, connect historical and cultural resources and life service facilities along the route, shape the important historical and cultural route of Nanjing actively, improve the quality of supporting facilities for

物质空间形态设计立足于三个层次：在城区层面，提升西五华里与老城南及东五华里的整体关系，建立特色路径，整合滨河区域历史文化资源；在地段层面，结合交通评价优化道路结构、传承历史街巷空间尺度、结合历史解读与商业策划明确功能能业态布局，并建立"虚拟地块"有效延续城市传统肌理；在滨河空间层面，建立公共开放、水陆交融、立体衔接的慢行系统，并基于对南京传统建筑要素特征的研究，塑造兼顾传承与创新的滨河风貌，提升空间环境品质与活力。

tourism and perfect service facilities for people's livelihood.

Design of material space and form is based on three levels: At the urban level, enhance the overall relationship between the "West Half Ten Miles" with south historical districts of Nanjing and the "East Half Ten Miles", build characteristic route to integrate the historical and cultural resources of riverside area; At the district level, combined with traffic evaluation to optimize the road structure, inherit the spatial scale of historical streets and lanes, combined with historical interpretation and business scheme, define the distribution of functions, and establish "Virtual Plots" to effectively maintain the traditional fabric of the city; At the riverside space level, establish a slow traffic system that is open to the public, easy to connect riverside and land, and connecting different levels. Based on the research of the characteristics of traditional Nanjing architecture, shape the riverside landscape with both inheritance and innovation, improve the quality and vitality of space environment.

P702

北京永定河城市滨水复兴新地标
New Landmark to Revitalize City Waterfront by Yongding River in Beijing

联合参展单位：筑境设计、北京市城市规划设计研究院、清华大学建筑设计研究院、北京清华同衡规划设计研究院、北京首钢建设投资有限公司、中国建筑设计研究院有限公司

联合参展人：施卫良、李兴钢、薄宏涛、鞠鹏艳、朱育帆、刘伯英、金洪利、周婷

Co-exhibitors (Companies): CCTN Architectural Design, Beijing Municipal Institute of City Planning & Design, Architectural Design & Research Institute of Tsinghua University, Beijing Tsinghua Tong Heng Urban Planning & Design Institute, Beijing Shougang Construction Investment Co., Ltd., China Architecture Design & Research Group

Co-exhibitors (Individuals): Shi Weiliang, Li Xinggang, Zhang Li, Bo Hongtao, Ju Pengyan, Zhu Yufan, Liu Boying, Jin Hongli, ZhouTing

永定河，北京市的母亲河，一直默默哺育着这座六百年古都。

作为永定河改造的重要段落之一的首钢园区沿岸，通过清理被工业园区堆场、煤场等侵占的滨水场地，将绿色空间还给城市，通过下沉埋地处理割裂岸线和园区的铁路货运线，缝合城市与滨水的空间联系，还之于城市。

滨水空间的缝合、再造与复兴

结合首钢园区的城市复兴进程，以强大的奥运IP引领城市更新为契机，沿永定河三公里岸线植入石景山公园、冬奥广场、三号高炉博物馆、国家体育总局冬训中心、香格里拉酒店、冬奥大跳台、三炼钢科创社区、耐火砖厂艺术中心等项目。基于永定河滨水城市休闲带的活力复兴，导入城市产业功能，提升生态空间品质，让滨水空间与城市同呼吸、共复兴。

Yongding River, the mother river of Beijing, has been quietly feeding this ancient capital for 600 years.

As one of the important sections of reconstruction of the Yongding River, the Shougang Park waterfront will return the green space to the city by clearing the waterfront sites occupied by storage yards and coal yards in the industrial park, and realize the spatial connection between the city and the waterfront and return it to the city by lowering and burying the railway freight line which separates the water front and the park.

Sewing, Reconstruction and Revitalization of Waterfront Space

It combines the urban revitalization process of Shougang Park, takes the powerful Olympic IP leading the urban regeneration as an opportunity, and introduces projects including Shijingshan Park, Winter Olympic Plaza, No. 3 Blast Furnace Museum, Winter Training Centre of General Administration of Sport of China, Shangri-La Hotel, Big Air of Winter Olympics, No. 3 Steel Works Science and Technology Community, and Refractory Brick Factory Art Centre which are placed along the three-kilometer coastline of Yongding River. Based on the vitality revitalization of the Yongding River waterfront urban leisure belt, the urban industrial functions will be introduced to enhance the quality of the ecological space, so that the waterfront space and the city will breathe together and revitalize together.

P703

黄向明团队
Team of Huang Xiangming

云形水石

Gift from Taihu — Cloud-shaped Rockery

苏州高新区文体中心位于浒光运河之滨，被高速路、山体和水道围绕，是对开山采石留下的城市生态"疤痕"的复绿利用。我们以多层次、多标高、多重院落的组合空间，建筑组群层叠退台和展开，以赢得更多的屋顶空间和攀爬体验，使17万㎡的超大体量消解为一个"隐形"的地景化建筑，成为与环境充分相容的、服务周边市民、还归城市的公共空间。顺势而成的云台之丘、数字像素化诠释的湖石，回应太湖山水及姑苏文化内核。

回归到建筑形体，"云形水石"的概念是文化与未来的延续：承袭苏州园林虚实结合的空间构型的同时，多意的弹性空间组织可以为未来运营提供更多可能。苏州的街坊和园林虚实相生和内外交替的通透，成为建筑"里中有外，外中有里"的基因。在这里，凹凸的体块使空间产生一种"留白"，形式与功能的对应不再单一不变，多意和辩证的孔洞结构的自主性可以通过复制和变异来适应环境，引导建筑自发生长而不破坏它的完整性。

Located on the shore of the Huguang Canal, the SND Cultural and Sports Centre is surrounded by highways, mountains and waterways. We landscape the remaining walls to treat the urban ecological "scars" left by the quarrying of mountains. The complex's multiple courtyards, varying heights, and a terraced form based on the site's topography bring more generous rooftop space and better climbing experience, so that the huge complex of 170,000m² can be merged into context as an "invisible" landscaping building and open spaces enable to host residents' daily activities. The complex which fit into hills and digitalized lake stone is in response to the Taihu Lake landscape and the core of Gusu culture.

As for the architectural form itself, the concept of "cloud-shaped rockery" carries on the city's tradition and symbolizes its future. In compliance with the unique spatial layout of classical gardens, the flexible spaces are designed to create further operational potential. The complex features the complementary relationship between its interior and the city's streets. The units of different volumes allow the form to respond to programs in various manners. The dependent structure riddled with holes is able to adapt to the context through duplication and variations, guiding the architecture to grow on its own without spoiling its entirety.

苏州高新区文体中心
The SND Cultural Sports Centre

P704

孟凡浩
Meng Fanhao

与日俱新，回应自然
Advance with Times, Echo with Nature

随着城市营建和乡村激活的不断深入，城市和乡村已成为不可分割的社会生态体系，需要建筑师以更宏观的视角介入与尝试。通过在城市和乡村的两个滨水项目的建筑实践，以不同策略回应自然、文脉以及时间性，尝试构筑特定环境下真正有价值的空间场所，追求人工与自然的和谐平衡，重新反思当下建筑师的责任与立场。

With the deepening practice of urban construction and rural activation, cities and villages have become an inseparable social and ecological system, which requires architects to intervene and try from a more macro perspective. Through the architectural practice of two waterfront projects in urban and rural areas, different strategies are adopted to respond to the nature, context and timeliness, to try to construct a truly valuable space in a specific environment, to pursue a harmonious balance between man-made and nature, and to rethink the responsibilities and positions of architects at present.

渔乡茶舍
Teahouse in Jiuxing Village

196

隐居江南精品酒店
Seclusion Jiangnan Boutique Hotel

P801

逐水而居——
青年建筑师新作
Living on Waterfront —
New Projects of
Young Architects

策展人：李翔宁、高长军
策展团队：江嘉玮、黄翰仪、李慧敏
参展人：范蓓蕾、范久江、何哲、孔锐、李丹锋、刘可南、戚山山、沈海恩、水雁飞、
苏亦奇、王硕、臧峰、翟文婷、张旭、周渐佳（按姓氏拼音排列）
Curators: Li Xiangning, Gao Changjun
Curatorial Team: Jiang Jiawei, Huang Hanyi, Li Huimin
Participants: Fan Beilei, Fan Jiujiang, He Zhe, Kong Rui, Li Danfeng, Liu Kenan,
Qi Shanshan, James Shen, Shui Yanfei, Su Yiqi, Wang Shuo, Zang Feng, Zhai Wenting,
Zhang Xu, Zhou Jianjia

锦溪大野之乐
Jinxi Lost Villa

上海之鱼景观桥
Footbridges in Shanghai Fish

较之于高山，滨水同样是建筑师心仪的建造基地。阿尔多·罗西在海滩灯塔的身上感受到孤独的"物感"，勒·柯布西耶在可眺望马丁岬岩礁的地方筑起隐修的"卡巴农"。在现代建筑的遗产里，滨水是一条能讲人性故事的涨落线。逐水而居，是这份遗产延续至当代的恒久命题。

这次展览组织了国内八家新锐青年建筑师事务所的滨水作品。从温润如玉的江浙沪区到粗犷壮阔的北域疆土，自然景观为建筑师备好了天然的素材。设计类型涉及民宿品牌、观景平台、基础设施等，这些作品通过现代建筑的基本语汇和组合来制造新的感官体验和亲水关系。建筑师在"逐水"议题之下，探讨了中国当代城乡与建筑的共同问题：既有表达预制体系特征的节点构造，也有提高乡村误差宽容度的在地建造，还有突显结构构件的设计技法。呈现在观众面前的八个新作品，以各自不同的方式回应了我们的未来生活可以如何与水发生关联。

Like high mountains, the waterfront has for long been a favored site by architects. From the lighthouse on the seashore, Aldo Rossi caught sight of the solitude of objects; above the cliff of Cap Martin, Le Corbusier built his hermitic cell Cabanon. The waterfront becomes a line of ebb and flow that narrates the stories of humanity in the legacy of modern architecture. It remains as a long-term issue for architects to think about how to live on waterfront.

This exhibition is curated upon the idea of promoting some recently completed projects on waterfront. Works by eight emerging young architect ateliers in China are chosen. Ranging from the warm and mild Shanghai-Jiangsu-Zhejiang Region to the vast and rough North China, natural landscape lays out the necessary ingredients for building. These projects cover related fields of hotel brands, landscape platform, infrastructure, etc., and all of them strive to construct a new relationship with the waterfront areas, and to create new sensual experience through a recombination of basic vocabularies of modern architecture. Under the theme of "Living on Waterfront", some common problems in contemporary Chinese urbanism/ruralism and architecture are explored: some of the projects focus on the articulation of prefabrication system by joineries, some focus on the enhancement of tolerance in rural construction, and some prefer to underline the existence of structural components by updating design strategies. These eight projects respectively respond to the future relationship of modern living and waterfront area in their own way.

临海 T 宅
House T

P901

雅克·费尔叶建筑事务所（法国）
Jacques Ferrier Architecture (France)

水边
On the Waterside

收集、运输和分享：水是城市生活的核心。在水龙头将我们与水的关系简单化之前，水井与喷泉将我们的社区生活组织得井井有条。水是社会的一部分，它为通讯和人类本身、知识与货物运输提供了可能性。然而，如今的城市似乎更多地承受着水带来的恐惧：洪水、污染、健康风险、卫生、饮用水、河流净化和地下水干涸等现象影响着城市生活，我们必须找回水的魅力。

Collecting, transporting and sharing water is central to urban life. Before the faucet isolated our relationship with water, wells and fountains organized the lives of communities, and water was part of society. It opened roads for communication and for the transport of people, knowledge and goods. However, cities today seem to suffer more from the fear of water: flooding, pollution, health risks, sanitation, drinking water, river purification and groundwater drying are affecting urban life, and we must find the charm of water again.

鲁昂诺曼底委员会
Metropole Rouen Normandie

P902

跨水越岸——连接汉堡港口新城
Crossing Water — Connecting the Hamburg HafenCity

gmp·冯·格康，
玛格及合伙人
建筑师事务所
gmp Architects
von Gerkan,
Marg and Partners

gmp·冯·格康，玛格及合伙人建筑师事务所（gmp）展示了三个位于汉堡港口新城的水岸基础设施项目，分别是 2018 年 12 月建成通车的汉堡易北河桥新地铁站，2018 年初建成的巴肯港自行车及步行桥，以及被列为汉堡州文物保护建筑、建成于 2002 年的仓储城 Kibbelsteg 桥。gmp 除了为人熟知的公共建筑、体育建筑、办公与会展建筑作品之外，在德国还有许多城市基础设施项目落成，其中最具代表性的是这些水岸项目。它们既是优秀的技术性解决方案，也是连接水岸的重要基础设施，更是体现了结构美学的作品。

展墙上最醒目的大幅手绘概念图，是 gmp 创始合伙人玛格教授，为港口新城做的前期研究而绘制的。1997 年汉堡规划部门以此为基础来制定进一步的规划。三个屏幕由左至右分别展示了 Kibbelsteg 桥、巴肯港自行车及步行桥及汉堡易北河桥新地铁站的项目视频。人们可在船只、海鸟与水声交织的海港背景音氛围中，身临其境地感受三个项目，也可通过展桌上的图纸资料与悬挂于桌面上方的 1:50 易北河桥新地铁站结构模型，可以了解更多项目信息。

gmp Architects von Gerkan, Marg and Partners (gmp) showed three waterfront infrastructure projects in the Hamburg HafenCity, namely the new Elbbrücken Underground station, which was opened to public in December 2018, and the bicycle and pedestrian bridge in Bakkenhafen, which was completed in early 2018, and the Kibbelsteg Bridge, built in 2002, listed as a heritage building in Hamburg. In addition to the well-known public buildings, sports buildings, office and congress & fair buildings, gmp have built many urban infrastructure projects in Germany, the most representative of which are these waterfront projects. They are excellent technical solutions, important infrastructures for connecting waterfronts, and works with structural aesthetics.

The most striking and large-scale concept sketch on the exhibition wall was drawn by Professor Marg, the founding partner of gmp, for the first study of the HafenCity. The study laid out many of the development principles which became the basis for further planning for the city authorities in 1997. The three screens show the project videos of the Kibbelsteg bridge, the bicycle and pedestrian bridge in Bakkenhafen and the new Elbbrücken Underground station from left to right. In the background sound with the atmosphere of the harbor, where the ships, seabirds and water are intertwined, visitors can experience three projects in an immersive way. More project information can be learned through the drawings on the table and the 1:50 structure model of the new Elbbrücken Underground Station hanging above.

P903

感性城市工作室（法国）
Sensual City Studio (France)

与城市一体
One with the City

《与城市一体》是一个通过虚拟现实技术拍摄的电影，让我们沉浸在纽约、马赛和上海，这三个有着极强身份的都市，她们代表着当今世界不同规划类型的城市。我们观察这三座城市的视角并不是从高处俯瞰，而是从它们的内部，通过当地人们的生活片段和氛围去探索。而这些被人居住着的空间可以让我们重新体验城市，重新思考感觉、情感、记忆和幻想在城市里的地位。我们通过影片询问着当今世界各地城市的规划进程。把观众放到体验的核心，通过这种沉浸的形式，迫使我们重新思考人体与城市环境的关系。

One with the City is a virtual reality film that immerges us in three different cities: New York, Marseille and Shanghai. We found that the perspective of these three cities is not from the top, but from the inside, through the local people's life segments and atmosphere to explore. And these inhabited spaces allow us to re-experience the city and rethink the place of feelings, emotions, memories and fantasies in the city. These three cities reflect different urban planning patterns. The film questions urban planning around the world today. Putting the audience at the heart of the experience makes us rethink the relationship between the human body and the urban environment.

P904

Let's Vote, 滨水活力评价展
Let's Vote, Vibrant Riverside Voting Exhibition

策展人：戴春

学术支持、系列论坛、活动策划团队：Let's Talk 学术论坛（金怡、董艳、董林涛、彭柏寒）

策展运营、展陈设计及布展团队：筑术空间（虞晓宸、王佳佳、吴薇、谭雅秋）

Curator: Dai Chun

Academic support, a series of forums, event planning team: Let's Talk Academic Forum (Crystal Jin, Dong Yan, Dong Lintao, Peng Baihan)

Exhibition operation, exhibition design and exhibition arrangement team: Archiepos Studio (Ignatius Yu, Wang Jiajia, Wu Wei, Tan Yaqiu)

本展区将展示 Let's Talk 对于滨水空间评价体系的研究。通过"拉洋片"、线上线下"投票"等互动形式，形象地让公众了解评价体系的意义和原理。引导大众参与滨江活力评估体系的共建，提出滨江空间设计后评估系统前置的可能性。

线下互动性展览区：墙面为评价体系介绍，附有投票球，地面呈现的是上海滨江地图，选取的滨江四个区域位置分别有展台展示其风貌，观众可通过手摇装置翻看其对应的动画图景。

线上投票：杨浦滨江沿岸和展览现场设置海报，观众可扫描二维码进入到线上评估系统，对滨江设计作品进行评价，展览从室内延伸到室外杨浦滨江。

This section will show Let's Talk's research on waterfront space evaluation system. Through interactive forms such as zoetrope and online and offline "voting", the public cultivate the aware of the meaning and principle of the evaluation system. This will guide the public to participate in the co-construction of the riverside vitality assessment system. At the same time, it will also put forward the possibility of the assessment system ahead of the riverside space design.

Offline interactive exhibition area: The wall concludes an introduction to the evaluation system, with a voting ball, a map of the Shanghai Riverside on the ground. The selected four riverside areas have their own booths to show their features. Viewers can view the corresponding animated pictures by hand-cranking devices.

Online voting: Posters are set up along Yangpu Waterfront and the exhibition site. Visitors can scan the QR code to enter the online evaluation system to evaluate riverside design works. The exhibition extends from indoor to outdoor along Yangpu Waterfront.

P905

上海刘滨谊景观规划设计有限公司
Liubinyi Landscape Planning and Design Co. Ltd., Shanghai

滨水景观
Waterfront Landscape

团队以刘滨谊教授为主帅,以同济大学为依托,以风景园林、景观规划设计、旅游策划规划为特色,具有近30年的景观规划设计学研究与实践积累。

在全国范围内,先后完成新疆喀纳斯湖、内蒙古成吉思汗陵、南京玄武湖、厦门鼓浪屿、浙江千岛湖、华山、洛阳龙门等国家级风景名胜区、国家级旅游区、国家级森林公园的策划与规划设计项目数十项;先后完成城市与景观综合性项目策划规划、城市绿地系统总体规划、城市森林总体规划、旅游区总体规划、各类产业园区、城市滨水区、城市街道广场、大学校园、旅游度假地、旅游宾馆、景区景点等总体与详细规划设计百余项。

主题阐述: 团队立足项目实际,开拓创新、倡导以景观为主,坚持景观、城市、建筑的三位一体,并且与市政、林业等多专业合作共事。以环保、绿色、低碳、节能等综合的环境与社会效益为首要追求目标,近年来参与完成国内多个滨水景观项目,从总体规划、详细规划以及方案至施工图设计,创造了多个滨水人居环境的典范。

The team is based on Professor Liu Binyi, Tongji University, and characterized by landscape architecture, landscape planning and tourism planning. It has accumulated the research and practice of landscape planning and design for nearly 30 years.

Since its establishment, the company has completed over hundred major projects in China, which include master plans & detailed plans of several national scenic areas, such as Kanas Lake in Xinjiang, Genghis Khan Cemetery in Inner Mongolia, Xuanwu Lake in Nanjing, Thousands Islands in Zhejiang and Hua Mountain in Shanxi. The company has also completed nearly one hundred projects in different areas of China, including investigating & evaluating landscape resources, master planning and design of industrial parks, urban green space system design, streets and squares landscape design, campuses landscape design, resorts design, and scenic areas landscape design.

The team is based on the actual situation of the project, pioneering and innovating, advocating landscape mainly, adhering to the tripartite of landscape, planning and architecture, and working with various disciplines such as municipal and forestry. Taking environmental protection, green, low-carbon, energy-saving and other comprehensive environmental and social benefits, the first priority is to pursue the goal. In recent years, it has participated in the completion of a number of coastal landscape projects in China. From the overall planning, detailed planning and the plan to the construction drawing design, it has created the model of multiple coastal living environment.

伊犁河
Yili River

暨阳湖
Jiyang Lake

P906

姜平
Jiang Ping

建构风景——融汇与对望
Architectonic Landscape — Convergence & Dialogue

我们近期的建筑实践聚焦跨越边界的多元建筑类型，尤其关注对于景观与建筑边界的探讨。亚洲高密度城市环境中的建筑通常因为空间的制约、景观介入的方式和程度受限而变得极为珍稀和关键。近几年的设计践行，包括张江人工智能岛、张江中区未来公园概念规划国际竞赛优胜方案，以及最近落成的西藏墨脱气象中心，在建筑与景观的边缘，尝试建筑化的风景与营造地景范式的建筑，自然植被、水体等景观元素与建构元素重构并交汇融合，呈现出多样化的介入方式与姿态，重新思考建筑本体与景观环境之间更多的可能性。

Our recent architectural practice has been focused on diverse architectural typologies across boundaries, with particular emphasis on exploration of boundaries between landscape and architecture. Interventions in high-density urban environments in Asia cities are often constrained and challenged due to space limitation. A selection of our recent projects, including the Zhangjiang Artificial Intelligence Island, Visitor & Sports Centre Future Park at ZJ Innopark, an international competition winning entry, and the newly completed Medog Meteorological Centre in Tibet, present a multifaceted perspective of intervention on the edge of the architecture and landscape. Natural vegetation, water bodies and other landscape elements are reconstructed and intertwined with the architectural elements, presenting diverse interventions and altitudes, and rethinking more possibilities between the architecture and the landscape environment.

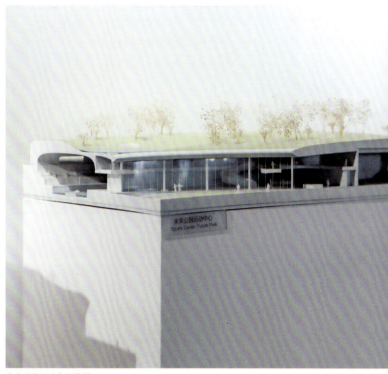

未来公园运动中心模型
Model of Sports Centre, Future Park

张江人工智能岛模型
Model of Zhangjiang AI Island

212

P907

张健 Zhang Jian

黄浦江两岸开发首个实质性项目——财富广场景观设计

Launching the First Substantive Project on Both Banks of the Huangpu River — Landscape Design of Fortune Plaza

该项目是被称为"百年大计，世纪精品"的黄浦江两岸综合开发的第一个实质性运作项目。设计突出上海作为国际化大都市的特色，以文化兼容来表现"海纳百川"的包容胸怀，熔地域性、文化性、时代性为一炉，传译船港文化的文脉，体现时代的气息。"临水而居，以水为财"，以水文化为焦点，创造性地运用了绿地覆盖、平台、景观墙等多种手法，艺术地处理了贯穿基地中心的防洪墙，设计了集港口历史、船舶知识和公共休闲于一体的绿色景观带，形成了一条人、绿、水、光交相辉映，自然之韵与人文景观交融的别具特色的国际滨水景观风景带。

This project is the first substantive operation project of the comprehensive development of both sides of the Huangpu River, known as the "Century Plan, Century Excellence". The design highlights the characteristics of Shanghai as an international metropolis, expresses the inclusive mind of "Haina Baichuan" with cultural compatibility, fuses regional, cultural and contemporary characteristics into one furnace, interprets the cultural context of the port, and reflects the flavor of the times. "Living near water, making money from water", focusing on water culture, creatively applied green space covering, platform, landscape wall and other means, artistically dealt with the flood control wall running through the centre of the base, designed a green landscape belt integrating port history, ship knowledge and public leisure, and formed a human, green and water landscape belt. The unique international waterfront landscape belt, which combines natural rhyme with humanistic landscape, is characterized by the radiance of lights and water.

214

福新第三面粉厂保护和再利用设计
Design for the Protection and Reuse of the Fuxin Third Flour Mill

福新第三面粉厂旧址为上海小沙渡底浜北，2009年因道路建设，原址往西北方向平移55米，现地址为光复西路145号。本次改造首先对现有建筑的格局进行评估，确定各部分构筑物的建造年代，以此作为位移时拆除的依据。其次，根据设计原则和保留建筑所在基地环境，确定建筑移位的最终位置和标高。同时在保存保留建筑实物本身的前提下，添加必要的建筑更新与安装工程。

The old site of Fuxin No. 3 Flour Mill is located in Bangbei, Xiaoshadudi, Shanghai. In 2009, due to road construction, the original site moved 55 meters northwest, and the present address is 145 Guangfu West Road. The reconstruction first evaluates the pattern of the existing buildings, determines the construction age of each part of the structure, as a basis for demolition when displacement occurs. Secondly, according to the design principles and the environment of the base where the building is located, the final location and elevation of the building displacement are determined. Finally, on the premise of retaining the physical object of the building itself, the necessary building renewal and installation works are added.

P909

回归的江河
The Return of River

上海市政工程设计研究总院（集团）有限公司景观规划设计研究院、钟律景感空间上海市劳模创新工作室

Shanghai Municipal Engineering Design Institute (Group) Co.,Ltd. Landscape & Planning Design Institute, Zhonglv Shanghai Labor Model Innovation Studio of Landscape Sense Space

理想的城市应该有一条理想的河流,它不是记忆中片断的河床,而是洁净美丽的栖所家园,河流被赋予它自己该有的全面意义,河流的综合设计在水体生态修复和生物栖地重建技术的基础上,融入城市公共游憩和文化创意设计,江河将重回健康和诗意的体验氛围,这是每一个城市都梦寐以求的画面。

从上海母亲河苏州河到黄浦江,从吴淞江到长江、湘江、珠江、赣江……我们的设计团队,通过对中国多条大江的滨水空间的技术实践和实地调研留下了对于专业发展的系统思考,并一直追求学科的跨界整合。

An ideal city should have an ideal river, which is not a river bed in memory, but a clean and beautiful habitat homeland. River is endowed with its own comprehensive significance. The comprehensive design of a river is based on the technology of water ecological restoration and habitat reconstruction, integrated into the urban public recreation and cultural creative design, which leads the river return to its healthy and poetic status. This is the dream picture of every city.

From Suzhou Creek-the mother river of Shanghai to Huangpu River, from Wusong River to Yangtze River, Xiangjiang River, Zhujiang River, Ganjiang River… Our design team, through the technical practice and field research on the waterfront space of many major rivers in China, left systematic thinking on professional development of rivers, and has been pursuing cross-border integration of disciplines.

P910

宅语・在水一方
Housing by Waterside

同济大学李振宇教授工作室
策展人：李振宇
执行策展人：卢汀滢、米兰
团队成员：王达仁、宋健健、
羊烨、王修悦、陈柳珺、
梅卿、王浩宇、张篪、田萌

Prof. Li Zhenyu Studio in Tongji University
Chief Curator: Li Zhenyu
Executive Curators: Lu Tingying, Mi Lan
Team Members: Wang Daren, Song Jianjian,
Yang Ye, Wang Xiuyue, Chen Liujun,
Mei Qing, Wang Haoyu, Zhang Chi,
Tian Meng

在滨水条件下，通过富有趣味的空间组织，以住宅建筑为载体，表达建筑内部空间与户外及至街区、城市的对答，为水景的观照创造特定的空间语汇。手语作为表达的出发点，在此时成为住宅建筑形式的新抽象。日常的及物性超越了其原有功能，成为设计工作的新语言。

By organizing interesting spaces and taking residential buildings as carriers, the residential architecture shows the relationship between the housing and the city at waterfront in both the interior and outdoor space. The exhibition creates a specific spatial vocabulary in response to the waterscape. As the basis of expression, gestures became a new abstraction of the form of housing. The daily transitivity expresses more than its original function and becomes the new language of design work.

P911

于洋
Yu Yang

人工的水
Water of Kunst

水是自然之物，但这个装置试图最大限度表现水的人工的一面：水是可以被自由引导的，水是可以被着色的，水是可以摆脱重力的。

Water is natural, but this installation attemps to show the artificial aspects of water: water could be canalized, water could be colourized, water could escape the limitation of gravity.

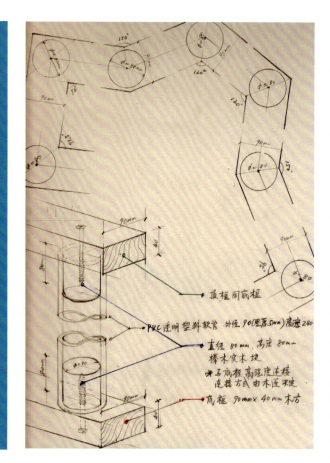

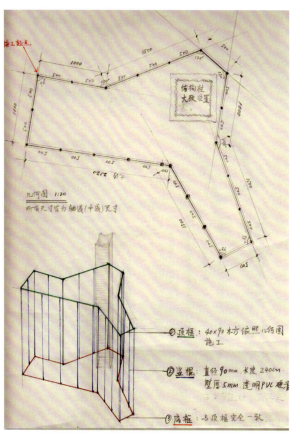

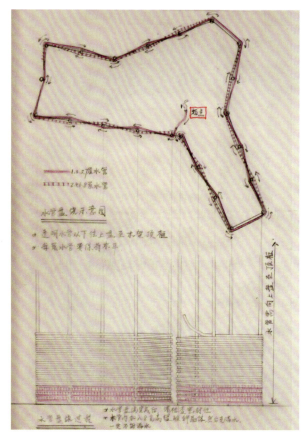

P912

水城印象
Overlooking the City with Water

AssBook 设计
食堂（贾涛、
闫泽明、侯之）
AssBook Design
(Jia Tao, Yan
Zeming, Hou Zhi)

关于上海这座城市，有太多关于"水"的记忆，我们希望通过这个专属的"城市之书"装置，向大众展现水与城市之间的密切关系，希望通过时间、地点、视角的转化，与大众产生新媒介互动交流。

"AssBook 设计食堂"是一家极具影响力的新媒体平台，致力于传播设计，让人们尊重并享受设计的价值。

"城事设计节"是由我们发起的大型城市更新活动。该活动连续三年在上海、深圳等地成功举办，现已成为国内极具影响力的城市更新领域策划执行主办方。

For Shanghai, there is too much memory about "Water", and we hope to show the close relationship between water and the city to the public through this exclusive "CITY BOOK". We hope that through the transformation of time, place and perspective, We will interact with the mass media to create new media.

"AssBook SheJiShiTang" is an influential new media platform dedicated to spreading design to respect and enjoy the value of design.

The "Urban Design Festival" is a large-scale urban renewal event initiated by us. This event has been successfully held in Shanghai, Shenzhen and other places for three consecutive years, and has become the influential host of the urban renewal field planning and execution

P913

舟游上海
Punting@Shanghai

○筑设计
Office ZHU

位于长江三角洲地区的上海，地势平坦，水网密布，历史上内河航运繁忙。然而，随着城市转型及填滨筑路开展，大量中心河道遭到废弃。对于这一议题，团队进行了一系列城市研究，并分别从历史、通勤两个层面入手，探讨上海中心城区河道潜力。展板主要由两种颜色叠加印刷而成，即洋红以及蓝色。透过红色、绿色滤镜，参观者可观察到上海中心城区内河不同层面的活动。通过丰富的画面及各色对话，展示魔都滨水河岸的多样性。

Located in the delta region, the downtown area of Shanghai used to have dense water network and busy shipping. However, with the urbanization and land reclamation, a large number of central rivers are covered or filled. To arouse the discussion of this topic, our team work on a series of urban studies. The potential of commuting and historical sightseeing along the Suzhou Creek have been explored during our study. The panel is drawn by two colours, magenta and blue. Through the red, green filters, visitors can observe two different layers activities (commuting & sightseeing) along the river bank. The rich line drawing and funny conversation also represent the metropolitan's vivid and magic life.

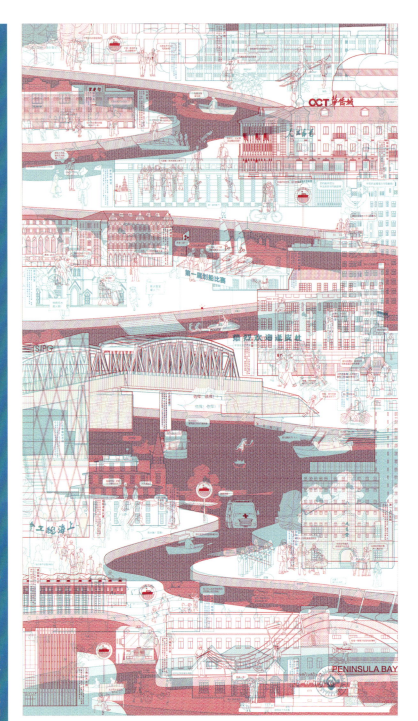

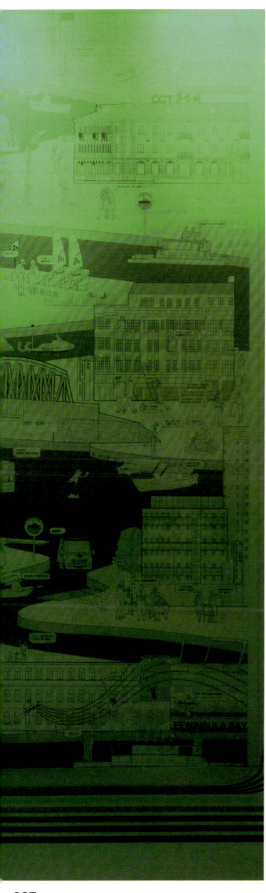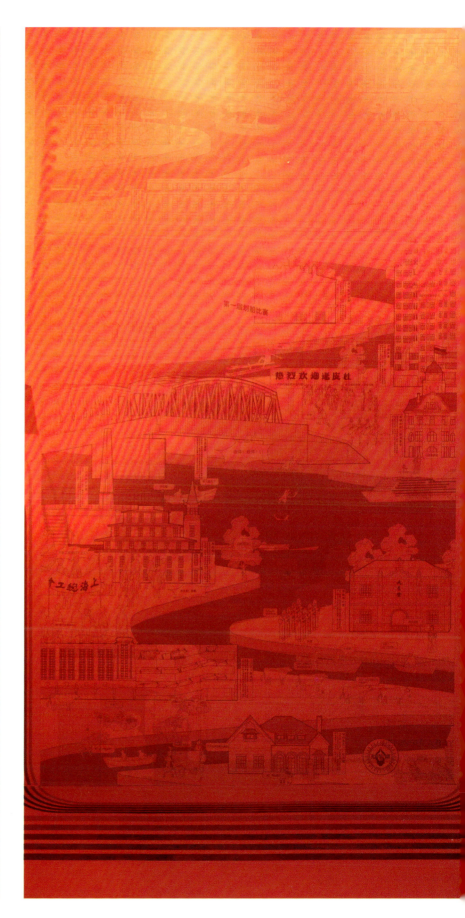

P914

吴建
Wu Jian

化学反应
Chemical Reaction

通过对东体育会路及其周边区域进行研究，将校园空间、生活社区、行政设施、自然要素、交通体系等不同资源进行梳理、分析与整合，通过对现有空间界面进行调整和重塑，使差异化的空间内容得以贯通与融合，并促成互动式、互补式、互助式发展模式的产生。由此引导"有机混合"与"共生发展"方式的生成，从而刺激有限的空间资源之间发生积极的"化学反应"，并哺育出新的内涵价值。这一过程突破了既有空间物理边界的桎梏，促使原有相互隔绝的空间要素之间产生更为积极的联系，在整体区域活化与共生的基础上，形成空间体系整体品质的提升。

Through the study of East Tiyuhui Road and its surrounding areas, we will sort out, analyze and integrate different resources such as campus space, living community, administrative facilities, natural elements and transportation system, and adjust and reshape the existing space interface. Differentiated spatial content can be integrated and integrated, and promote the development of interactive, complementary and mutual development models. This leads to the formation of "organic hybrid" and "symbiotic development" methods, which stimulates positive "chemical reactions" between limited spatial resources and fosters new intrinsic value. This process breaks through the ambiguity of the physical boundary of the existing space, and promotes a more positive connection between the original isolated spatial elements. On the basis of the activation and symbiosis of the whole region, the overall quality of the space system is improved.

城像—成像

P915

姚栋
Yao Dong

可复制、可推广的农民相对集中居住设计模式

A Duplicable Mode of Centralized Peasant Housing Design

新江南田园，不仅要美得超群，还要可复制可推广。现今上海乡村的农村住宅不仅因在形式上追求"洋气"而丧失了地域风貌，更因以城市为标准的建设使其整体环境失去了郊野的韵味。本案以漕泾镇水库村农民集中居住点一期南片为例，提出了"强化自然禀赋，生产性公共空间，传统轮廓未来内涵"三项设计策略，打造了一个体现传统村落原真美、游业相宜、面向未来的新江南田园，探索了一条可复制可推广的农民相对集中居住设计模式。

The new form of Jiangnan villages should be beautiful and also easy to be duplicated and emulated. Today's village houses of Shanghai have lost their identities due to the pursuit of foreign trends while the essence of being villages is fading because they are trying to be like cities. Taking the centralized peasant housing planning and architecture design of Shuiku village, Caojing as an example, this project comes up with 3 ideas: enhance the nature talent, make public spaces productive and give the vernacular architecture new power. These ideas will help to build authentic and new Jiangnan villages which combine tourism industry and agriculture together. Moreover it is making efforts to find a duplicable mode of centralized peasant housing design.

P916

山东临沂蒙山大洼
艺术谷滨水公共空间再造
The Regeneration of Waterfront
Public Space in Mengshan Art
Valley at Shandong Linyi

同济大学建筑与城市
规划学院（宋玮、田唯佳）
Tongji University College of
Architecture and Urban Planning
(Song Wei, Tian Weijia)

在当代乡村振兴语境中，"公共空间"通常以一种变异的形式所存在着，公共空间所服务的对象不再是本地村民，更多的是消费乡村空间的异地游客。本项目尝试打破传统的本地—异地二元对立的格局，以"水"为载体，对场地现有的"乡村大舞台"进行二次设计，将服务于游客的酒店配套设施开放给当地村民，使其拥有真正意义上的公共空间属性。项目中的景观步道重新连接原本孤立的各个功能场所，将滨水而生的新旧公共空间重新整合，从而达到消除村民与游客的活动边界，重建乡村滨水空间公共氛围的目的。

In the recent context of rural revitalization, "public space" usually has been treating as a variant form. The public space serves not only local villagers but also tourists from the city who actually are consuming the rural space as a way of experiencing the nature lifestyle. The exhibition shows a project aiming to break the traditional conflict opposition and pattern between local and outside, and uses "water" as the carrier to renew the existing "The Grand Stage" on the site, so as to open the private facilities serving only for tourists to the local villagers, in order to make those places have the real public space attribute. The walk way in the project reconnects the originally isolated sites and integrates the old and new public spaces along the waterfront, so as to eliminate the activity boundary between villagers and tourists and to rebuild the public atmosphere in the rural waterfront space.

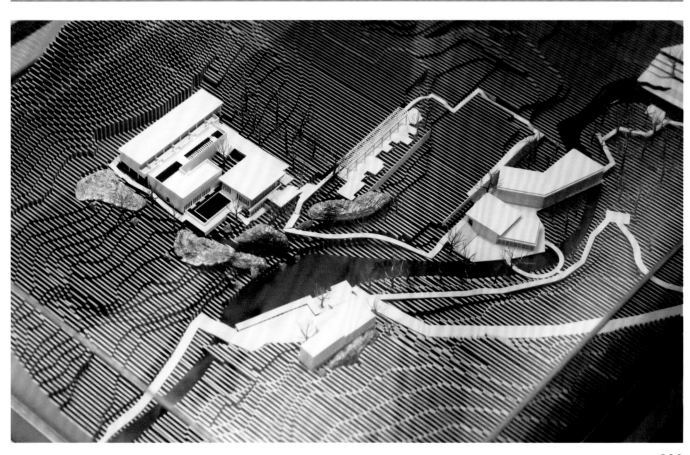

P917

朱祥明／上海市园林设计
研究总院有限公司

Zhu Xiangming / Shanghai
Landscape Architecture
Design&Recearch institute
Co., Ltd.

延续历史文脉的滨水空间
重塑——福建泉州五里桥
文化公园案例

Reshaping of Waterfront
Space Inheriting Historical
Context –Quanzhou
Wuliqiao Culture Park, Fujian

城市，是沿着水系而建立发展的，择水而居；

城市的发展都是由岁月书写而成的，岁月留痕；

滨水空间是城市中最复杂的生态地段之一，敏感脆弱。

"我们只是地球上的旅游者，来去匆匆，但城市是要永远存在下去的。"——贝聿铭

本案集建筑遗产保护、生态环境恢复、游憩空间塑造于一体：一座以历史文化遗产与生态环境恢复为主要特征的生态文化公园。

五里桥具有 870 多年的历史，是我国现存最长的梁式石桥，具有极高的历史文化价值。由于经济的高速发展、管理的滞后，五里桥周边区域存在违章建设、环境污染、生态退化、社会安全等诸多问题，桥体也受到越来越严重的破坏。本案从建筑文化遗产保护、周边自然环境修复、市民游憩空间拓展 " 三位一体 " 的体系出发，不仅着眼于对桥体本身的保护，同时也注重对其周围生态自然环境的恢复与治理，特别是水体生态环境的修复，为市民提供良好的生态游憩环境。

The city is established and developed along the water system and live by water;

The development of the city is written by the years, leaving traces of the years;

Waterfront space is one of the most complex ecological sections in the city, which is sensitive and fragile.

"We are just tourists on earth, coming and going in a hurry, but the city is going to last forever. " (Ieoh Ming Pei)

This case sets the trinity of architectural heritage protection, ecological environment restoration and recreation space shaping: an ecological cultural park with the main characteristics of historical and cultural heritage and ecological environment restoration.

Wuliqiao Bridge, the longest ancient beam stone bridge widely known all the world with more than 870years history, has extremely high value in civil and historical aspects. During the recent decades, the surrounding areas near the bridge have been influenced by illegal constructions, environmental pollution, ecological degradation and social safety due to rapid economic development and management stagnation. The bridge has suffered from these factors and has become worse in recent years. In this study, the trinity methods had been established for the protection of architectural culture heritage, natural environmental restoration and enlargement of recreation space. To achieve this aim, not only the protection of the bridge itself but the ecological restoration and management of surrounding environment recovery were concerned, especially the ecological restoration of water environment. The design obtained from this study could be helpful to effectively provide the good ecological recreation space for residents.

P918

胡沂佳、饶广禛
Hu Yijia, Rao Guangzhen

屏的风景——
永恒自然中稍纵即逝的瞬间
The Scenery of the Screen — A Fleeting
Moment in the Eternal Nature

展项以 E+Lab 事务所设计的永宁驿站为案例，通过城市 mini 综合体介入滨水空间，激活在地空间活力。

作品以"屏"作为空间组织中的核心语言，以竖向条纹的镜面不锈钢所映射的风景为媒材，重新开启与自然风景的一种新的关系和体验，以建筑装置的方式捕捉、再现与凝聚风景，发现隐藏在风景图像之下的色彩密码。

The exhibition takes the example of Yongning Station, designed by E+Lab, which activates the waterfront space's vitality by a Mini urban complex.

The design uses "screen" as the core language, and uses mirrored stainless steel with vertical stripes as a media to reopen a new relationship between the experience and nature, to capture, reproduce, and condense landscapes in the installation way and to discover the colour passwords hidden under landscape images.

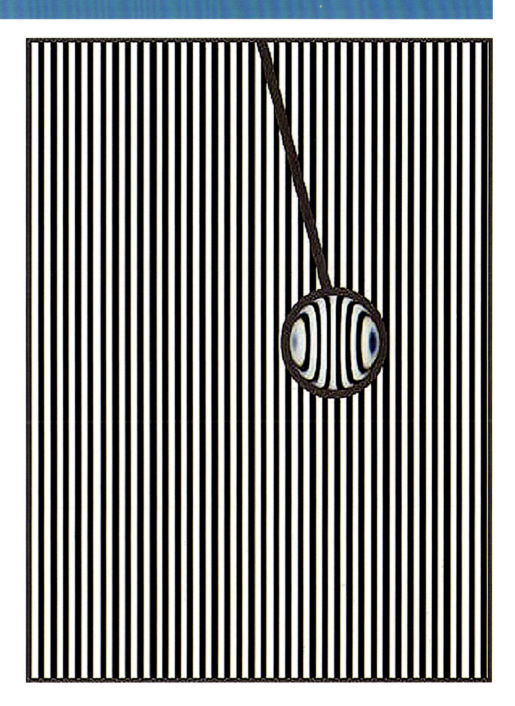

P919

无限环梯
Infinite Slide

贾涛
Jia Tao

无限环梯的设计灵感来源于数学模型中的莫比乌斯带。而无限环梯的设计初衷便是将 3 个莫比乌斯带通过数字化设计手段，巧妙地结合面与面、线与线之间的关系，使其成为一个整合儿童滑梯与儿童攀爬壁功能的景观都市家具。上海黄浦江沿岸的滨江区域是一块不可多得的集景观、人文于一体的都市滨水空间。将无限环梯这样一个具有儿童游乐设施功能的都市家具置于滨江，可更近一步提升市民生活品质，体现本届上海城市空间艺术季——滨水空间为人类带来美好生活的全球性基调。展览时将呈现详细的设计图纸和分析图、效果图及 3D 打印模型。

The design of the Infinite Slide was inspired by the Mobius Belt in the mathematical model. The original design of the Infinite Slide is to combine the three Mobius belts ingeniously through parametric design method. And through the combination of faces, it has evolved into a kind of urban furniture, which integrating children's slides and children's climbing wall function. The riverside area of Huangpu River in Shanghai is a rare urban waterfront space that integrates landscape and humanity. Placing the Infinite Slide, an urban furniture with the function of children's play facilities, on the waterfront can further enhance the quality of life of the citizens and reflect the global keynote of SUSAS 2019 - waterfront space brings a better life to human beings. The exhibition will be presented by detailed design drawings, renderings, diagrams, and 3D printed models.

无限环梯效果图

Rendering of the Infinite Slide

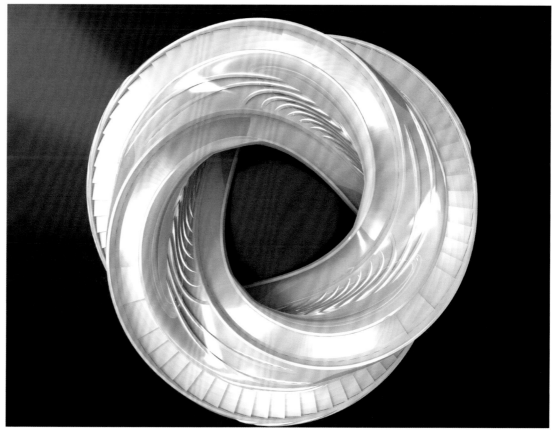

P920

松园
Song Yuan

郭文／未相景观与城市设计事务所
Atelier VISION

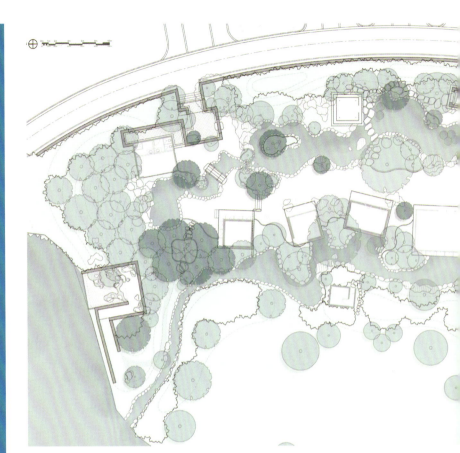

王羲之的《兰亭序》描绘了千年前一场临水而娱的上巳节雅集，他以山水之美和欢娱之情，抒发对于生死无常的感慨。"松园"的造园概念正是来自于兰亭雅集。设计不满足于仅为业主设计一个喝茶场所，而是希望创造一个寄情山水的文会之地。设计的布局沿一条水系展开，区分了内、外环境，外部公共绿地被场地西侧水系隔开，内部主景区域的户外活动沿水岸展开，茶室依水布置。设计探讨了人与水的关系，思考了园林在当代的呈现以及扮演的角色。

The Orchid Pavilion — a poem written by notable scholar Wang Xizhi, describes the celebration of events along the water during the Shangsi festival in ancient China. Contemplating the beauty of nature, the mountains and the joy of entertainment, Wang Xizhi expressed his feelings on life and death. The Song Yuan project is inspired by the written artwork of The Orchid Pavilion. We propose to transcend the mere concept of a tea house by creating a cultural landscape which encourages us to both physically and spiritually connect with the space. The layout of the Song Yuan proposal unfolds along the winding canal, which creates a threshold between the internal and external environment. The outdoor public green space is separated by the water system on the western side. Arranged along the water, activities in the main landscape area are merged with the tea houses to generate internal spaces. The design further aims to explore the potentialities of the relationships between people and water, while also considering the representation and role of Chinese gardens.

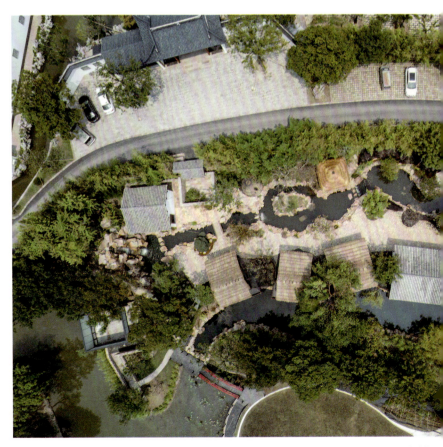

空间艺术版块
Urban Space Art

在杨浦滨江岸线上极具象征性的历史遗迹——拥有近百年历史的上海船厂船坞和毛麻仓库，由中外艺术家刘建华、高桥启祐、彩虹合唱团、毕蓉蓉等创作装置、影像和音乐作品，以综合性的艺术手法表达"船坞记忆"和"水的相遇"；在毛麻仓库二层，管怀宾、杨牧石、胡为一、程然、何翔宇、邱加等16位中国当代艺术家的作品相遇百年历史的毛麻仓库，通过捕捉并理解上海的过去和现今，创作他们的当代艺术作品，借上海16个行政区的数字，构成"上海16景"。进入建筑空间本身，就是对上海这座城市在时间维度的纵深的一种探索，16种观点则提示着上海的多元风景和丰富内涵。上海不仅是表达的对象，也是点燃灵感的火光，它给生活和停留在这里的人们以记忆，也给未曾造访的人们以幻想。上海的光景构成人们的一部分，而人成为这座城市最生机勃勃的风景。

The nearly century-old Shanghai Shipyard Docks and Maoma Warehouse are the symbolic historical sites on the Yangpu riverfront. Here, artists from China and abroad including Liu Jianhua, Keisuke Takahashi, Shanghai Rainbow Chamber Singers, and Bi Rongrong create installation, video and music works on the theme of "Memory of Shipyard" and "Water Encounter". On the second floor of the Maoma Warehouse, 16 Chinese contemporary artists, including Guan Huaibin, Yang Mushi, Hu Weiyi, Cheng Ran, He Xiangyu, and Qiu Jia, create their contemporary artworks on the basis of Shanghai's past and present. Their artworks together constitute the program of "16 Sceneries of Shanghai", named after the fact that Shanghai is comprised of 16 administrative districts. Entering the architectural space itself is an exploration of the depth of the city in the time dimension, and 16 viewpoints suggest the diverse scenery and rich connotation of Shanghai. Shanghai is not only the object of expression but also the fire of inspiration. It gives memories to those who live and stay here as well as fantasies to those who have not visited. The scenery of Shanghai forms a part of the people, and people become the most vibrant scenery in the city.

A001

封存的记忆
Sealed Memories

刘建华
Liu Jianhua

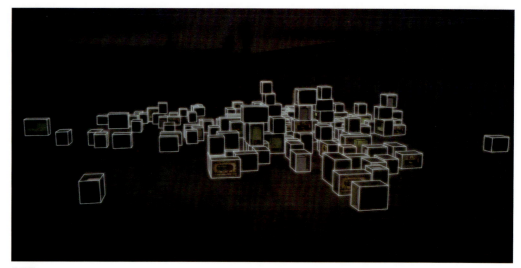

效果图
design sketch

作品利用船坞遗址的机器来呈现、梳理一个历史的空间。杨浦滨江造船工业从清末时期开始，一直持续到 21 世纪初。从更深层次思考，这里的造船工业对中国社会形态的发展也有极大的推动作用。用工业化的材料钢板将这些不同的机器封存，而有些又是敞开的，从中通过一些视频看到 20 世纪 80 年代中国社会在当时是如何进行反思及拥抱未来的。我们今天的一切都与那个时代密切相关，而上海这个港口城市也是其中的一个缩影，如果淡忘 80 年代就意味着我们是否还能前行……作品呈现的方式在外形上相对极简和抽象，白天观众可以通过一些影像及改造后的装置了解历史发展的线索，晚上这些装置安置的冷光带又会带给观众不同的视觉变幻的效果，让人在观看艺术的同时也产生对历史的敬意。

The work is made up by the machines left in the ruin of a shipyard, with the intention to show a historical space. Shipbuilding at Yangpu can be trace back to the late Qing Dynasty and survived until the early 21st century. It also played a great role in promoting the development of the Chinese society. The artist uses steel plates to seal up these different machines while leaving others open where visitors can see how China in the 1980s introspected itself and embraced the future. Everything we have today is closely related to that era, and the port city of Shanghai is also a microcosm of it. If we forget the 1980s, we may not move on. The work presents itself in a simple and abstract way. In the daytime, the audience can grasp the historic clues by watching videos and transformed objects. At night, the luminescent objects will bring visual changes to the audience, and arouse the respect for history among them while they view the art work.

A002

水之记忆
Memory of Water

高桥启祐
Keisuke Takahashi

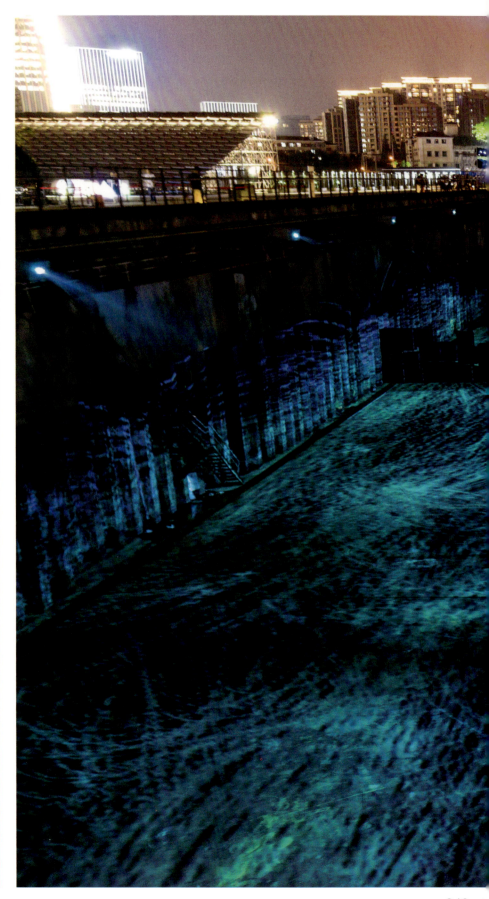

"河川总是与住在那里的人们共同存在，支撑着他们的生活。我想黄浦江也是这样。我通过这部作品，探索城市的过去、未来，以及新世界的形态。那里有遗失的风景和故事。"
——高桥启祐

作品描绘了被水分隔开的地形和风景的平缓界线，以及连接现在、过去、未来的时间变化的分界线，将整个空间作为河流和城市的缩影，将世界以微型化的形态呈现。黄浦江是这座城市生命的源动力之一，而这里的造船厂则是曾支撑一个时代的地方。艺术家试图描绘在已过去的时间里，人们心中最初的光。

"Rivers always exist with the people living there and support their lives. I think the Huangpu River is the same. Through this work, I explored the past and future of the city and the form of the new world. There are lost scenery and as many stories as the scenery." — Takahashi Keisuke

This work depicts the gentle boundary between topography and scenery separated by water, and the boundary connecting the changes of present, past and future. It regards the whole space as a miniature of rivers and cities and presents the world in a miniature form. This river has always been the motive power supporting the life of this city, and this shipyard can be said to be the place supporting the times. Takahashi Keisuke would like to describe the initial light in people's hearts in the distant time.

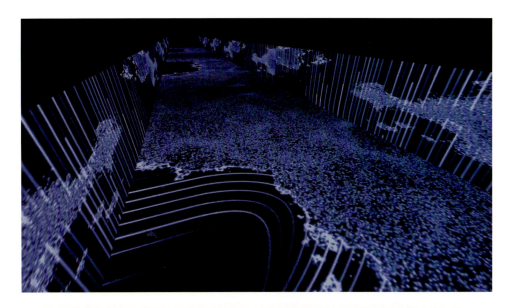

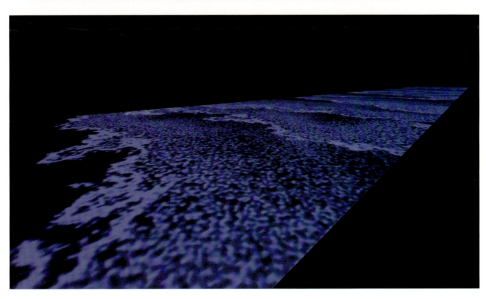

A003

彩虹合唱团
Rainbow Chamber Singers

合唱套曲《相遇》
Song Cycle *Encounter*

《相遇》是上海彩虹室内合唱团受上海城市艺术季委约，于 2019 年完成的艺术套曲，由艺术总监金承志完成词曲创作，音乐家田汩、罗赓、吴经纬共同完成音乐制作及编配。专辑由七首单曲组成，按时序描绘了杨浦滨江的七个片段，有江边的电厂、江边的人、人和人的故事、永恒奔流的江水等。其既是一天之中的七个缩影，也代表着杨浦滨江的前世今生。

不同于之前的套曲创作偏向室内音乐的特点，《相遇》在创作和制作过程中，在部分章节中尝试性地大量融入了当代音乐的创作技法和音响元素，尝试以更为丰富的音乐手法向听众传递更多层次的听觉感受，进一步尝试拓展合唱艺术的可能性。

1 遥远的呼唤

2 奔腾的江水

3 高耸入云

4 卖花姑娘（沪语口白：舞台剧《繁花》主要演员）

5 繁星落江

6 水底世界

7 故事

Commissioned by this year's Shanghai Urban Space Art Season, Shanghai Rainbow Chamber Singers (RCS) has written a song cycle *Encounter*. It is a joint work by the group's artistic director Jin Chengzhi, also the piece's songwriter, and Tian Mi, Luo Geng and Wu Jingwei, the musicians responsible for its sound engineering. The album is consisted of 7 singles, each depicting a particular scene of the Riverside of Yangpu District in chronological order. Together these scenes demonstrate the seven different states the river is in during a day, and how it flows from the past to the present.

In composing *Encounter*, RCS has swung from its usual predilection for chamber music features to contemporary music features, in the process incorporating the creative techniques and the audio elements of contemporary music into some parts of the song cycle in an effort to furnish the audience with a more enriched and layered listening experience, and further explore the possibilities of chorus art.

1 A Call from Afar

2 Roiling River

3 Piercing the Sky

4 Flower-selling Girl (with dialogues in Shanghai dialect by lead actors and production team of the drama *Blossoms*)

5 Stars Fall on the River

6 Underwater World

7 Stories

A101

毕蓉蓉
Bi Rongrong

流动之物
Things that Flow

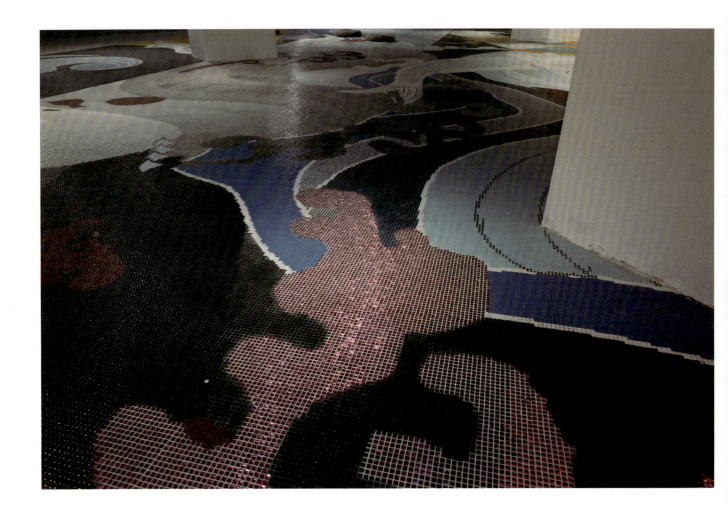

艺术家毕蓉蓉近几年通过旅行不断收集来自不同时间和空间的纹样。她通过绘画、织物、场域特定的装置、动画，对这些纹样碎片进行再诠释。这也是她对历史、对现实环境的一种回应方式。

本次展出的作品中，她将目光投向上海城市兴起过程中扮演重要角色的江河与海水，采集了中国传统山水画、石刻、古籍《山海经》中的水纹样或水中生命的纹样，以及在旅途中所拍摄的现实中的水形象，将这些纹样组合拼贴。毕蓉蓉创作了一幅铺在地面的巨大水纹图景，与悬浮在空间之中、流动于时间之中的水波动画形成互文。

结合动画与场域，毕蓉蓉邀请了艺术家徐程为这件作品创作了声音部分。徐程以多年采集的各地水声来充盈整个空间，融化的雪水、湖中浪潮、江南细雨都汇集于此，纯声声景从细节质感的层面再现水之变化。

In recent years' journey, artist Bi Rongrong has been collecting patterns from the different time and space. She re-interpreted these pattern fragments through her paintings, fabrics, field-specific installations, animations. This is also a way for her to respond to history and the real environment.

In this exhibition, she paid more attention on the river and the sea that play an important role in the rise of Shanghai. She collected the patterns of water and life from the traditional Chinese landscape painting, stone carvings, and the ancient book *The Classic of Mountains and Seas*, as well as the real water images taken during the journey, and combined these patterns together. She created a huge water pattern on the ground, and intertexture with the water wave animation suspended in space and flowing in time.

Combine animation with field, Bi Rongrong invited the artist, Xu Cheng, to create a sound part for the work. Xu Cheng filled the whole space with the collected water sound for many years. The sound of melted snow water, the lake wave and the Jiangnan rain were all gathered here. The acousmatic scene reproduced the change of water from the level of detail.

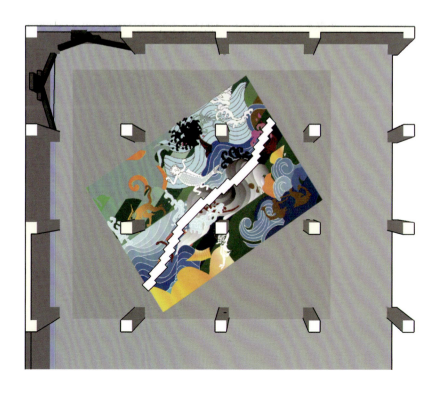

A102

殷漪
Yin Yi

向西
Toward the West

《向西》是以苏州河上的四川路桥为场所创作的8声道5频的声音影像装置。

在作品中，一位音乐家在城市的一座桥上表演音乐，这位音乐家在与不同时间中的自己合奏。通过向城市注入音乐，艺术家殷漪用音乐把城市中不同时间、不同空间的声音和影像编制起来，营造出浅吟式的城市场所。邀请观众在展览现场用自己的身体和意识去构建在时间空间重叠中溢出的四川路桥。

《向西》延续了近年来艺术家殷漪对于"声音动力影像"的创作与探索。作为上海出生长大的市民，他也在用艺术方式提出自己对于这座城市的观察、思考与评价。

殷漪邀请艺术家照骏园为《向西》创作四重奏《四个时间》，同时照骏园也在作品中出演音乐家这一角色。

Toward the West is a 5-channel quadraphonic audio-video installation created in the location of Sichuan Road Bridge on Suzhou Creek.

The installation shows a musician playing an orchestra on a bridge of Shanghai. The band consists of various personages of the same performer at various moments of the city.

Bringing music into the city, Yin Yi interlaces the sounds and images of different times and spaces in Shanghai, building a place that croons. The visitors are invited to engage their bodies and minds with a Sichuan Road Bridge that overflows from an overlapped time-space.

Toward the West is a continuation of Yin Yi's art work and exploration on "sound-momentum picture" in recent years. As one born and growing in Shanghai, he provides his observation, contemplation and evaluation of the city with his art.

Yin Yi invites Zhao Junyuan to compose an instrumental quartet *Four Times*, the instrumental quartet in *Toward the West*, is composed by Zhao Junyuan who is also musician.

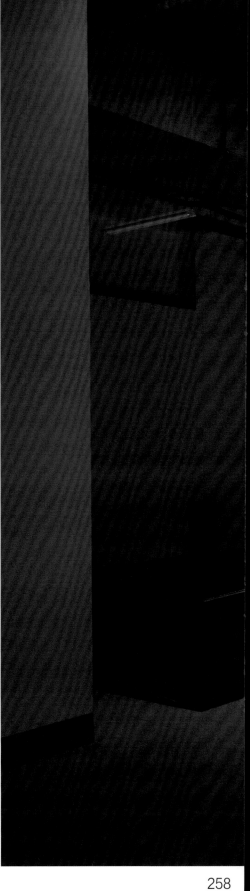

A201

外部世界
Umwelt

施政
Shi Zheng

《外部世界》是一个由四个屏幕组成的影像装置，其中主要视觉部分采样于上海崇明岛的东平国家森林公园。艺术家通过数字技术将崇明岛独特的岛屿自然景观重建，在从"自然"至"科技"的转译中呈现一种"超真实"的景观。

同时，连接并支撑四个屏幕的支架结构作为自然中"树木"的隐喻，支撑着生态系统中物质与能量的不断交换与传递。在这平行的"仿真"现实中，作品试图通过对工业化都市中的森林进行"复制"和再现，从而探讨在当下受人类影响的科技时代中"自然"的真实性的问题。

Umwelt is a video installation composed of four screens, the main visual part of which is sampled in Dongping National Forest Park in Chongming Island, Shanghai. The artist uses digital technologies to reconstruct the unique natural landscape of the Chongming Island and present an "ultra-reality" of the landscape by translating the "natural" to the "science and technology".

At the same time, the support structure of four screens is connected and supported as a metaphor of natural "tree". It supports the continuous exchange and transmission of material and energy in the ecosystem. In this parallel simulation, the works attempt to "copy" and represent the steel forest of industrialized cities in order to explore the authenticity of "nature" in the era of science and technology influenced by human beings at present.

A202

张如怡
Zhang Ruyi

闪电、银块、带电的石块
Lightning, Silver Block, Electric Gravel

张如怡的创作以个体存在与现实环境变迁之下的隐秘关系为主要方向，她善于观察，捕捉关于空间、物质、个人内在意识等之间的相互作用力，并以美学的方式物化。

艺术家以上海杨浦区的工业风景为灵感，借助既有空间中的方柱结构，运用水泥和瓷砖的墙面分割场域，配合混凝土的植物造型的雕塑、电线、浴室地漏以及视频等不同媒介，进行视觉语言的交互，制造出一个被工业化之后的日常物证。

Zhang Ruyi's works mainly focus on the hidden relationship between individual existence and the change of real environment. She is good at capturing the interactions between space, material and individual's inner consciousness, and materializing it in aesthetics.

Using the interaction between industrial scenery and daily life, the artist uses existing square pillars and cement/ceramic tile wallsto divide the space. And through with different media such as sculpture, wire, bathroom floor drain and video, she manages to create a witness to what daily life looks like after industrialization in an interactive visual language.

A203

事实是，我很容易陷入爱恋
The Truth is that
I Fall in Love, So Easily

郝经芳、王令杰
Hao Jingfang, Wang Lingjie

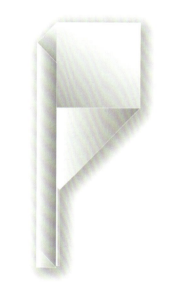

A MAPLE SEEDS ORIGAMI MACHINE — HAO & WANG

CEILING

A MACHINE
SILVER METAL
&
TRANCEPARENT

↳ A PAPER ON THE FLOOR

竖直的装置从地面拾取纸张并向上传送，几个机械结构逐步将纸片折叠成类似枫树翅果的形状，而后纸张被吹离机器装置，缓缓下落，划出一道漂亮的螺旋轨线。经过几秒的时间，折纸降落在地上。而在这之间，新的一张纸又被机器向上传送，开始新的折叠、下降的过程，周而复始。

作品的灵感之一来自艺术家年幼时一次学校出游，城市里长大的艺术家第一次在上海青浦见到以不可思议的螺旋姿态下落的枫树种子。这件作品仿佛是多年以后艺术家对童年在大自然所见到风景的回应。

The vertical installation picks up the paper from the ground. Then mechanical structures fold the paper into a shape similar to that of maple samara. Eventually, the paper is blown away from the machine, drawing a beautiful spiral line and landing slowly on the ground. In a few seconds, another piece of paper will be picked up, folded, blown before falling and landing. Up, Up and down.

One of the Inspirations of the work comes from one of the artist's childhood trips to Qingpu, Shanghai, where the artist first saw maple seeds falling in a spiral posture. In this way, the work is a response to a childhood scene.

A204

相遇上海
Encounter Shanghai

施海
Shi Hai

海派、时尚、摩登、风度、礼仪、优雅、睿智，这就是我眼中的上海和上海人。

艺术家施海用巨大的绅士礼帽象征上海的文明精神，而上海礼帽与白玉兰的相遇，构成了艺术家心中的上海风景。

Shanghai style, fashionable, modern, graceful, polite, elegant and wise, this is what I think of Shanghai and Shanghai people.

Shi Hai symbolizes the civility of Shanghai with a huge gentleman's hat, while the combination of Shanghai hat and magnolia white constitutes the scenery of Shanghai in the artist's mind.

A205

光的背面
The Dark Side of Light

冯晨
Feng Chen

《光的背面》是一件场域特定装置作品。本次展览中，作品被安置在黄埔江边，透过声音控制的百叶窗，可以看到黄浦江以及对岸的风景。

艺术家通过现实采集到的声音分别控制百叶窗的开阖，此起彼伏，好像百叶窗在面对着黄浦江窃窃私语。

作品对透进展厅内的自然光进行改造和调控，使静止的空间有了光线带来的动感和韵律。

The Dark Side of Light is a localized installaztion specific to Huangpu River. Through the louvers controlled by sound, you can see the scenery of the river and its beyond.

The artist controls the opening and closing of the shutters with recorded sounds, as if the shutters were whispering in the face of the Huangpu River.

The work transforms and regulates natural light in the exhibition hall, adding dynamics and rhythm of light to a static space.

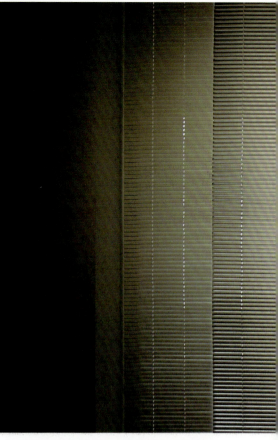

A206

邱加
Qiu Jia

等待命名的景观
To be Named:
2019.001/D, 2019.004/D,
2017.027/V

"作为一个土生土长的上海人，我确信自己并不会将上海以一种客体化的方式来感知。因为我一直处于其中。"

邱加试图将上海的"景"，还原到"位置"这一属性，最终成为一个意义的聚合体的当下——从上海的现实生活中提取切片的物质信息，如在上海被限制、淘汰、废弃的旧式木制家具，在一种对抗的逻辑中，将之打散、重置、组合，并在空间中重新放置，使"物"本身成为位置。艺术家通过作品创造的"位置"，使"逗留于此位置"构成无数种连接的可能性，并生成意义的聚合。

"As a native Shanghainese, I am sure I will not perceive Shanghai in an objectified way, because I have always been in it."

Qiu Jia tries to restore the "landscape" of Shanghai to "location" which eventually becomes an aggregate of meanings by extracting material information from the real life of Shanghai, such as the old wooden furniture which was restricted, eliminated and abandoned in Shanghai, and smashing, resetting, assembling and emptying them in the logic of confrontation and repositioned in space, so that the "thing" itself becomes a position. Through the "location" created by the artist's work, "staying in this location" constitutes a connection of countless possibilities and a convergence of meanings.

A207

海平线
Sea Horizon

管怀宾
Guan Huaibin

"海平线"是海天交合的地理界域，也是视觉溯望的消逝线；是城市的背景，也是家园的前沿；是它启程的起点和归属地，也是光与黑暗、明灭与流动的亲历者。

"海平线"沐浴着潮起潮落的波澜，也是长夜破晓启明的瞬间与落日余晖转承再启的长韵。

以《海平线》为题，源自艺术家管怀宾对上海依江临海独特地理位置的感怀，也与百年来上海与外部世界的种种交集以及快速的城市化进程所引发的一系列文化景观相关。两支铜质圆环的表面，蚀刻着几十组与光和黑暗相关的中英文词汇，并穿梭在正反 20 个金属小屋之间。观众在这里既可以凭栏远望、溢发想象；也可以触碰这个充满能量的界域，体验光和黑暗的精神内核，感悟风云跌荡的海上世界与上海城市景观的关联。

"Sea Horizon" is the geographical boundary between sea and sky, as well as the vanishing line of visual retrospect. It is the background of the city and the forefront of the home. It is the starting point and home of the other side's journey, also the witness of light and darkness, twinkle and flow.

"Sea Horizon" is bathed in the waves of ebb and flow of tide, but also the dawn of the long night and sunset moment turned into a long rhyme.

Sea Horizon is the artist Guan Huaibin's impression of Shanghai's unique geographical location along the Huangpu River and by the East Sea. It is also related to the convergence between Shanghai and the world through the past century and a series of cultural landscapes triggered by rapid urbanization. The surfaces of the two copper rings are etched with dozens of English and Chinese vocabularies related to light and darkness, shuttle between the front and back of 20 metal huts. The audience can look into the distance and overflow imagination, touch this energetic boundary, experience the spiritual core of light and darkness, and learn about the relationship between the turbulent sea world and Shanghai's urban landscape.

A208

17区
District 17

程然
Cheng Ran

微缩模型构成的雕塑内部放映的短电影，是艺术家虚构的"未来上海的风景"。微缩景观内仿真的岩石、湖水和瀑布，映射着由柔性屏幕发出的光泽。程然用其超现实观感的影像，以虚构的上海第17区来呈现一段发生在想象中的超级城市——未来上海的个人化的故事片段。

The short film, which is made of miniature and played inside the sculpture, is an artist's fictional "future scenery of Shanghai". The simulated rocks, lakes and waterfalls in the miniature landscape reflect the luster from the flexible screen. Cheng Ran uses his surreal videos to present a personalized story fragment in the fictitious th district of Shanghai, an imaginary super city in the future.

A209

上藩市
Shang Francisco

杨圆圆
Yang Yuanyuan

自2018年以来，杨圆圆开始对20世纪海外粤剧戏台、电影片场与夜总会场景中的华裔女性展开研究，并开始电影拍摄的旅程。在创作过程中，其在旧金山、夏威夷、纽约、哈瓦那等地拍摄素材，并进行了大量有关海外华人群体的历史研究。本次展出的影像作品《上藩市》属于该项目的一部分。

《上藩市》讲述了两段发生在旧金山的华人故事。在"上海来的女士"中，20世纪40年代生于上海、50年代搬来旧金山的Ceecee在家迎来了七年未造访的妈妈。101岁的母亲总是在口中喃喃自语："这是哪里？我在上海吗？"Ceecee将自己的家族故事娓娓道来后，也讲述了她与在上海的前夫跨越语言与地理限制，在千禧年通过交友网站结缘的爱情故事。"中国城轶事"则由一场旧金山中国城步行之旅展开：随着40年代奥森·威尔斯的电影《上海来的女士》走入旧金山中国城最后一家现存的戏院，又从"上海楼"漫步到"紫禁城夜总会"，镜头跟随着华裔舞者Cynthia的步伐，踏上一场时空交叠的旅程。

Since 2018, Yang Yuanyuan has been researching the Chinese female on the overseas Cantonese Opera stage, in movie studios and nightclubs, hence embarking on the journey of film shooting. In the process, she went to San Francisco, Hawaii, New York and Havana, etc. for the material and conducted a lot of historical investigations on overseas Chinese communities. The video *Shang Francisco* is part of this project.

Shang Francisco tells two stories about the Chinese in San Francisco. In "The Lady from Shanghai", Ceecee, who was born in Shanghai in the 1940s and moved to San Francisco in the 50s, welcomes her long-separated mother in her house, with her mom mumbling," Where am I? Am I in Shanghai?". After her well-told family history, Ceecee speaks about how she met her Shanghai ex-husband in spite of language difference and physical distance on a dating website in 2000. Anecdotes of Chinatown started with a walking tour in Chinatown, San Francisco. In the 40s, as Orson Wales' movie *The Lady from Shanghai* was shown in the last subsistent cinema in the Chinatown, San Francisco, then from the "Shanghai House" to "the Nightclub of the Forbidden City", the camera followed the ethnic Chinese dancer Cythina's steps upon the journey of overlapping space-time.

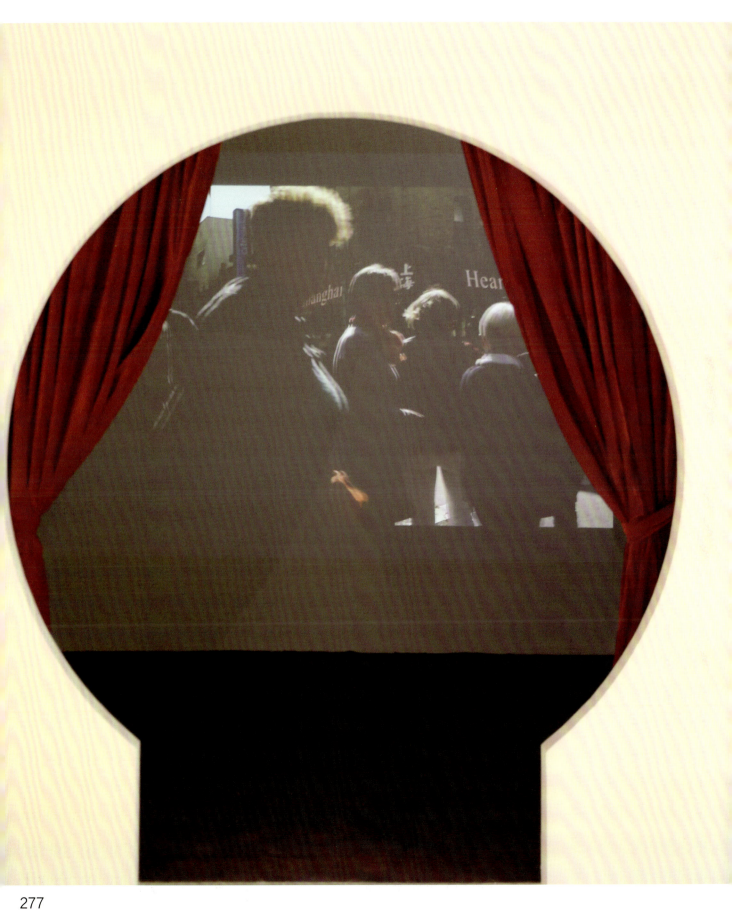

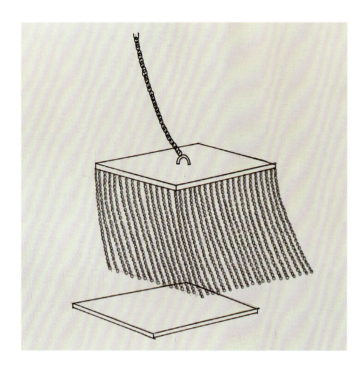

艺术家刘诗园以上海"租界"的"界"字为关键词，用珠链制作了四块悬挂在空中的方形空间。穿过这些扑面而来的体积，进入到灰色空间界内时，空间会随观众的进出而轻轻摇摆，并伴随着拨动珠链的声音。与动作相比，此处的声音是延迟的，在进入到内部空间之后，声音还在持续。珠链正下方的瓷砖，是艺术家对上海法租界里的教堂、寺庙、法式建筑以及梧桐树下的老房子的感受的视觉表达。

作品名为《雨匆匆，打错门》，像一个诗意的场景。然而凡是雨水必沾染灰尘，艺术家试图通过拨开珠链，拨开华美事物的表面认识其背后的真实。瓷砖图案中的一条白色的丝带，仿佛要将人们从阴差阳错的"打错门"，到达"对"的地方。

The artist, taking "limit" in "Rented Limited Area" (commonly known as Concession) as the keyword and creates four square realms with beads. When the visitors enter into the gray space limits, the volume will directly come onto your face. The space will sway gently as the visitors moves in and out, leaving the voice of moving beads eched in the space. Compared to the action, the sound is delayed and continues after entering the inner space.The ceramic tile directly below the bead chain is the artist's visual expression of the churches, temples, French architecture and the old houses under the plane trees in the French concession of Shanghai.

The work is titled *Wrong Rain, Wrong Violence*, which is like a poetic scene. However, the rain is always stained with dust, and the artist tries to push aside the chain to explore the truth behind the surface of the beautiful things. A white ribbon in tile's pattern leads people from "wrong door" to the "right" place accidentally.

A211

无痕、轻取、风的规范
No Traces, Easy Obtainment, The Rule of the Wind

胡为一
Hu Weiyi

艺术家胡为一关注飞速城市化进程中的城郊景观和住民的生活状态，在红砖墙围起的半开放空间里呈现的三件影像装置作品，是艺术家捕捉的三个场景。

《风的规范》里，空无一人的柏油马路上，只有风在遵循着信号灯的指令。《无痕》里，原本空荡荡的建筑，逐渐被生活气息填满，城市里柔软钢筋水泥的是生活的风景。《轻取》里的万家灯火回应着屏幕外灯光的呼吸而点亮、熄灭。

生活在其之中，却又跳离其外，艺术家向城市这个复杂的生命体投以冷静而透彻的目光。

What the artist Hu Weiyi focuses on is the suburban landscape and residents' living conditions amid the escalating urbanization process, and the three video installations exhibited in the semi-open space surrounded by red bricks and residual walls are three scenes the artist captured.

In *The Rule of the Wind*, only the wind is following the instructions of the signal lights on the empty asphalt road.In the *No Trace*, only the empty building, originally full with vitality, stands there, after the brush has gone by, without a trace behind. Myriad lights in the *Lightly Take* gradually go out as the street lights, broken off to the ground, go on. In the Code of Wind, only the wind follows the order of signal lights on the empty bituminous street.

Living inside, but escaping outsides, the artist views the city -- the complex with calmness and clearness.

A212

聚集——组

Aggregating — Group

杨牧石
Yang Mushi

八件线形结构的黑色雕像在空间中切割，与白盒子环境形成对比。生硬的转折面，不稳定的透视，矛盾的对称关系与比例，精密的连接点，过度打磨的痕迹在聚合的体量下发酵变异。艺术家杨牧石把在上海所收集的二手木质墙板、工作台、梯子和木地板切割后重新拼接，并在喷漆后塑造成八件向空间放射的体块。空间的尺寸大小参考了城市的室内居住空间和工作场所，它们呈现出被挤压的状态。黑色的雕像作为城市躯干的象征物，与展厅建筑的构成元素产生对话。艺术家通过对上海的日常用品的处理以及空间的改变制造出一个可读的、对城市进行提示的集合体。他试图以此实践激发观者对于城市发展以及生产关系的思考与想象。

Eight black linear sculptures cut the space, and contrast with the environment in white boxes. The rigid turning surface, unstable perspective, conflicted symmetric relation and scale, precise connection points and over-polished traces develop and change under the polymerized mass. Yang Mushi cuts and reassembles the second-hand wood wall panels, workbenches, ladders and wood floors that he collected in Shanghai, paints and shapes them into eight blocks radiating into the space. The space's dimensions refer to how large a place urban denizens are restricted to as they live or work: squeeze. The black sculptures, as the symbol of urban trunk, produce a correlation with the component elements of structures in the exhibition hall. The artist has created a readable aggregation indicating the city through processing daily supplies in Shanghai and changing the space. With this practice, he tries to inspire the viewers to think about urban development and production relation at present and imagine what lies in the future.

A213

Study: 亚洲男孩
Study of Asian Boy

何翔宇
He Xiangyu

这一雕塑生动捕捉了一位小男孩打开可口可乐易拉罐的瞬间，在传达情感与亲密关系之上的无间断张力的同时，亦对文化挪用进行再次繁殖。它既表达了西方文化进入东方所衍生的犹疑与好奇，也一并传达着"开启"这一动作所生发的主动性的、蓄势待发的力量。而将可乐再次激活的动作本身也在于思考其作为象征的本质及再生的意义，并探究符号意识的形成方式。在触及自我与他者在内容产出与意义输出上的模糊界限的同时，也通过"自我"这一媒介与他者间的联系对社会产生影响。

另一组平面作品则以海报的形式与雕塑共同呈现了可口可乐与上海的历史联结与文化承袭。这件雕塑与这组平面作品以综合表现形式深度呈现了可口可乐的企业文化，以及该品牌与上海这一国际化大都市之间源远流长的联结。它既是对储存的历史证据的创新性尝试，也表达了原发的唯美等同于欲望这一观点。

This sculpture vividly captures the moment when a young boy opens a Coca-Cola can, conveying the unrelenting tension of emotion and intimacy while reproducing cultural appropriation. It expresses the hesitation and curiosity that comes with the entry of Western culture into the East but also conveys the proactive, poised power of the act of "opening". The act itself is also a reflection on the nature of Coke can as a symbol and its meaning of regeneration, and an inquiry into the way symbolic consciousness is formed. While touching on the blurred line between one's self and others in terms of content and meaning output, it also has an impact on society through the connection between the "self" and others.

Another group of graphic works uses the form of posters and sculptures to present the historical connection and cultural inheritance between Coca-Cola and Shanghai. This sculpture and this group of graphic works deeply display Coca-Cola's corporate culture and the long-standing connection between the brand and Shanghai. It is not only an innovative attempt to store historical evidence, but also expresses the view that original aesthetics is equivalent to desire.

A214

刘毅
Liu Yi

红移
Red-shift

"生活在上海的异乡人，像海上的一块浮萍，在虚构与真实的生活中不断寻找自我。每个渺小的自己，努力与这个城市发生融合，反应。城市新移民与原社区人民形成一种理性兼容合作的新二元关系。"艺术家刘毅聚焦生活在上海的异乡人。灯箱以流水状悬挂在空间里，每一只灯箱里叠合6张不同的水墨画，流动的笔触相叠，构成一个瞬间的移动。这场移动可能是一次历史性的迁徙，也可能是游客般的短暂停留。

灯箱内的感应装置，使灯箱在观众经过时产生明灭。"每一组灯箱画，就像是一个流动的历史，你的靠近，唤起一段瞬间移动的历史，你的走动也成为这里的记忆，无论是永居还是短暂的停留，如电磁辐射的波段，在可见光波段里，普线朝红端移动了一段距离。"

"Strangers living in Shanghai are like pieces of duckweed on the sea. They constantly seek identity in life, virtual or real. Each neglible individual strives to integrate into and react with this city. The relationship between new migrants and old citizens is also changing, and establishes a new duality in place of mutual isolation, rejection and opposition." Liu focuses on the strangers living in Shanghai. The hung light box is in the shape of flowing water, and 6 pieces of different paintings are overlapped in each light box to constitute a momentary movement.

Embedded sensors make the light boxes twinkle when someone passes by. "Each group of light box picture is like a flowing history. It arouses a momentary and flowing history when the visitor walks around and gets close to it. Living and staying are like electromagnetic waves: in the band of visible light, the spectral line moves a bit towards RED."

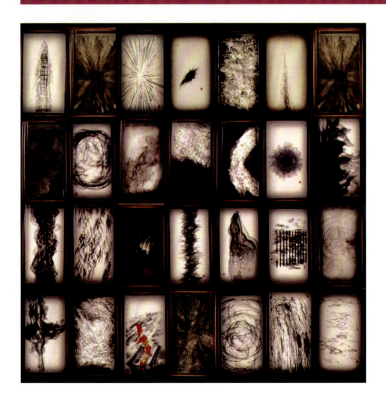

A215

袁松
Yuan Song

风景组合——上海大世界
View Combination — The Great World

始建于1917年的"上海大世界"曾是民国时期上海最吸引市民的娱乐场所。重新开业后，曾经的娱乐项目也被现代的技术和材料重新包装。

艺术家袁松使用如今现实生活中可以象征"大世界"历史印象的材料创作的"风景组合"：光鲜与破旧的对比、灵巧转动的机械、刺激的视觉假象。作品营造出一种既相互冲突又整体统一的次序关系。看似荒诞的材料组合一如艺术家对在当今这个消费时代中生活的人们的日常状态的感受——真实与幻想、矛盾与共存，发展上升，极具活力。

"The Great World" established in 1917 used to be the most attractive entertainment venue of Shanghai during the period of Republic of China. Now that it is reopened, old shows come back with modern technologies and materials.

"View Combination" is created by Yuan Song with the materials which symbolize its history. The work creates a stressful oneness by comparing freshness and oldness, flexibly-rotated machinery and stimulating visual illusion. The seemingly absurd material combination reflects the artist's vision about how ordinary people live in this age of consumerism—reality and fantasy, contradiction and coexistence. It is a work of vitality and force.

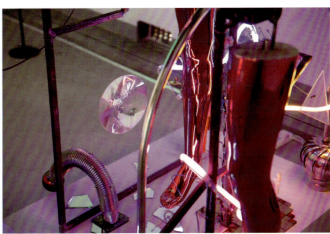

A216

中心说
Centre's Talk

汤杰
Tang Jie

艺术家汤杰将金属椭圆片按大小秩序悬挂排列，中间大两侧小，构成悬浮于空中的双锥体形态。

观者靠近时，感应装置将控制中心的最大圆片左右旋转，中心处圆片的旋转运动将能量依次传递至相邻的其他圆片，并发出声响。

装置的运动过程就像是上海这座城市向周围世界发出信号、传递能量与动力的过程，将抽象的文化现象以具象的方式呈现。而作品与观者的互动性提示着，人的行为是能量与动力传递的重要一环。

Tang Jie arranges a number of hung metal elliptical plate in such a way that larger ones are placed in the middle and smaller ones on the sides, constituting a double cone in the air.

Whenever a viewer gets close, the sensors will make the largest plate in the centre to move, taking other plates with it and hence transferring momentums it holds while specific sounds are made.

The movement indicates that Shanghai is emitting signals and energies to the world, presenting an abstract cultural phenomenon in a concrete manner. The interaction between the work and the viewers hint that human is important in transferring energy and momentum.

杨浦滨江公共空间
Yangpu Waterfront Public Space

杨浦滨江是从秦皇岛路到黎平路的沿黄浦江滨江带，绵延 5.5 公里，临江而立的是大量 20 世纪 40 年代以前建造的工厂、仓库和市政基础设施。这里的岸线记载了上海近代以来的工业化、城市化历程。在杨浦滨江从"工业锈带"到"生活秀带"的转型过程中，2019 年上海城市空间艺术季的举办是重要节点。

虽然艺术季的展期只有两个月，却完成了一次对杨浦滨江公共空间从形式到内容的全面动员。艺术季触达的不仅有上海船厂船坞、毛麻仓库的主展场，更为岸线留下了 20 件来自国内外知名艺术家的公共艺术作品，杨浦滨江的建筑与景观也成了艺术季期间活动发生的重要空间载体：气势宏大的开幕式在船坞中举行，沿岸线的设计师快闪活动为了解杨浦滨江的历史与设计意图打开了窗口，"城市的野生""共生构架"下的乐高积木搭建等活动使市民的参与永久地改变城市景观……

为此，城市空间艺术季绘制了这份杨浦滨江公共空间地图，从公共艺术、当代设计、建筑遗产三个角度加以介绍。它不仅是一份地理空间上的导览，有着对景点的详细介绍，也是对艺术季如何介入城市、触达生活的生动记录。

Start from Qinhuangdao Road to Liping Road, with thriving factories, warehouses and municipal infrastructure built before 1940s, Yangpu Waterfront, a strip of 5.5km along Huangpu River, made considerable contributions to the industrialization and urbanization of modern Shanghai. In the transformation from a "Post-industrial Belt" to "Lifestyle Show Belt", it is an important node to hold Shanghai Urban Space Art Season (SUSAS 2019) at Yangpu Waterfront.

With only two months of exhibition period, SUSAS 2019 has actually completed a comprehensive mobilization of Yangpu Waterfront public space from form to content. During the art season, not only the main exhibition sites of Dock of Shanghai Shipyard Co., Ltd. and Maoma Warehouse, but also 20 public artworks from well-known domestic and foreign artists were left along the waterfront. The architecture and landscape along Yangpu Waterfront have also become important activity spaces during the art season: the grand opening ceremony was held in the dock; the Flash Mobs of designers along the waterfront opened a window to understand the history and design intention of Yangpu Waterfront; and the activities of "Wildness growing up in the city" and LEGO building under the "Symbiosis Frame" enabled the participation of citizens to permanently change the urban landscape...

The Urban Space Art Season has drawn up this map of Yangpu Waterfront public spaces, which is introduced from the perspectives of public art, contemporary design and architectural heritage. It is not only a geographical guide with detailed description of sites, but also a vivid record of how the Art Season interacts with the city and touches the city life.

2019 上海城市空间艺术季
展场介绍
2019 Shanghai Urban Space Art Season Main Exhibition Area and Site

主展场——杨浦滨江南段（5.5 公里）

城市滨江地带通航便捷、景色优美，具有得天独厚的公共活动空间和景观资源优势。黄浦江从淀山湖迤逦而来，一路吸纳江南水乡的条条水路，走过青浦、松江、奉贤、闵行、徐汇、浦东、黄浦、虹口、杨浦、宝山十个区，江水连着城市的血脉，是上海密集的城市建设区内宝贵的公共空间走廊。一个世纪以来，黄浦江见证了上海从一个小渔村到现代化大都市的成长，也见证了上海近代工商业的发展。浦江两岸曾经厂房、码头、仓库林立，很长一段时间以来，只有外滩是人们可以亲近浦江的地方，在其他的岸线段，人们听得到汽笛的鸣响却很难见到江水的波涛。

2010 年世博会开启了黄浦江岸线功能转型的序幕，也标志着上海的城市发展进入越来越注重城市品质建设的阶段，决心要用美好的空间来回应人民对美好生活的期盼。

2017 年 12 月 15 日，上海 2035 规划获国务院批复原则同意。规划明确，黄浦江两岸地区将重点发展创意设计、博物博览、传媒等功能，成为世界级滨水文化功能带。几乎与此同时，2017 年 12 月底，在市委、市政府的重要部署和沿线各单位持续不断的努力下，黄浦江两岸终于实现了 45 公里岸线贯通。这一公共空间建设的伟大壮举响应了国家和人民对黄浦江转型发展再出发的期盼，不仅为上海市民提供了景观优美的活动空间，为滨江地区城市更新发起了开端，也以其深刻的人文关怀为世界城市建设史增添了亮丽的乐章。

2019 年 1 月 31 日，上海市政府出台了《关于提升黄浦江、苏州河沿岸地区规划建设工作的指导意见》，批复了《黄浦江沿岸地区建设规划（2018—2035 年）》。这两个纲领性文件进一步明确将黄浦江规划定位为全球城市发展能级的集中展示区，其中，杨浦滨江规划定位为未来的科创中心。

杨浦滨江，位于黄浦江岸线东段，滨江岸线长 15.5 公里，呈"拥江抱海"蓄势待发之势。杨浦滨江所在的杨树浦工业区为上海乃至近代中国最大的能源供给和工业基地，百年以来，作为上海近代工业的起点，代表了上海城市工业文明特色，在城市经济和社会生活中发挥着巨大的作用，被称为"中国近代工业文

Main Exhibition — Southern Section of Yangpu Waterfront (5.5km)

With convenient navigation and beautiful scenery, the urban waterfront has unique advantages of public activity spaces and landscape resources. Huangpu River meandering from Dianshan Lake absorbs all the waterways of water towns in the south of the Yangtze River, passing through ten districts of Qingpu, Songjiang, Fengxian, Minhang, Xuhui, Pudong, Huangpu, Hongkou, Yangpu and Baoshan. The river connecting the blood vessels of the city is a valuable public space corridor in the dense urban construction area of Shanghai. Over the past century, Huangpu River has witnessed how Shanghai evolved from a small fishing village to a modern metropolitan in the course of a century, and how a modern industrial and business world has developed here. The banks of Huangpu River used to be a forest of plants, wharfs and warehouses. For a long time, the Bund was the only section where people could get close to Huangpu River, while wherever else the tides were overwhelmed by the steam whistling.

Thanks to the EXPO 2010, the functions of Huangpu banks began to change, which marked a new stage of Shanghai urban development characterized by increasingly emphasis on urban qualities and determined responses to the citizens' prospect of lovely life with beautiful spaces.

On 15 December 2017, Shanghai 2035 Plan was approved by the State Council in principle. It is clearly planned that those areas on both sides of Huangpu River will be developed with the focus of creative design, exhibitions, media and other functions, and become a world-class cultural waterfront belt. Almost at the same time, the end of December 2017, by virtue of the significant arrangements made by Shanghai Municipal Committee of CPC and Shanghai Municipal People's Government as well as the continuing efforts of all involved units, the 45km Huangpu Waterfront Connection was finally completed. This great feat of public space construction responds to the nation and people's expectations for the transformation and development of Huangpu River. The project not only offers scenic activity spaces for the citizens, initiates the urban renewal of waterfront area, but alos adds a brilliant chapter to the global history of city construction with its deep humane concerns.

On 31 January 2019, Shanghai Municipal People's Government introduced the *Guiding Opinions on Promoting the Planning and Constructions Along Huangpu River and Suzhou Creek* and officially approved the *Construction Plan Along the Banks of Huangpu River (2018-2035)*. This two programmatic documents further confirm the target of Huangpu River banks as a concentrated exhibition area for the developmental capacity level of a global city, among which, Yangpu Waterfront is positioned as the future science and technology innovation centre.

Yangpu Waterfront, located in the eastern section of Huangpu River shoreline, is 15.5 kilometers long, with the shape of embracing both the river and the sea. Yangshupu Industrial Zone, where Yangpu Waterfront is located, is the largest energy supply and industrial base in Shanghai and even modern China. For one hundred years, as the starting point of modern industry in Shanghai, it has represented the industrial civilization of Shanghai and played a huge role in the economical and social life of the city, so called as "Corridor of Modern Industrial Civilization in China". In 1869, the public concession authorities built Yangshupu Road on the former riverbank of Pujiang River, which opened the prelude to the century-old industrial civilization of Yangshupu area; by 1937, there were 57 foreign factories and 301 national industries in the industrial zone; by the first half of

明长廊"。1869 年，公共租界当局在原浦江江堤上修筑杨树浦路，拉开了杨树浦百年工业文明的序幕；至 1937 年，工业区内已有 57 家外商工厂，民族工业已发展到 301 家；至 20 世纪上半叶已发展成为中国近代最大的工业基地之一，创造了中国工业史上无数个工业之最。

杨浦滨江的城市更新再发展，规划将分为南、中、北三段进行。其中南段西起秦皇岛路、东至定海路、南临黄浦江岸线、北至杨树浦路，岸线约 5.5 公里，滨江核心区域约 1.8 平方公里。在过去的 20 年中，杨浦滨江南段率先经历了工业仓储功能外迁、规划调整、土地收储、分段实施公共空间建设等阶段，开始由工业生产岸线逐渐向城市生活岸线转变。2019 上海城市空间艺术季的展场，即位于杨浦滨江的南段岸线滨水第一层面建筑至黄浦江之间，已实现贯通开放的公共空间区域。

杨浦滨江的工业遗存规模宏大，分布集中，具有较高的历史文化价值、科学技术价值和艺术审美价值。在杨浦滨江南段，规划建议保留历史建筑总计 24 处，共 66 幢，总建筑面积 26.2 万平方米，除此之外还保护保留了大量极具特色的工业遗存设施设备。这些工业遗存中有不少曾是中国工业史上非常重要的建筑，如：

远东最大火力发电厂——杨树浦发电厂（1913 年）；

中国最早的自来水厂——杨树浦水厂（1883 年）；

近代最长的钢结构船坞式厂房——慎昌洋行杨树浦工厂间（1921 年）；

中国最早的钢筋混凝土结构的厂房——怡和纱厂锯齿屋顶的纺车间（1911 年）；

中国最早的钢结构多层厂房——江边电站 1 号锅炉间（1913 年）；

近代最高钢框架结构厂房——江边电站 5 号锅炉间（1938 年）等。

滨江连绵的工业区曾经是横亘于黄浦江和城市生活空间之间的"隔离墙"。工程打开了封闭的生产空间，变成开放的生活空间，彻底改变"临江不见江"的城市空间状态，实现了"还江于民"。同时，贯通工程在空间景观设计中，致力于挖掘呈现工业遗存的文脉底蕴与场所记忆，寻求将工业区原有的特色空间和场所特质重新融入到城市生活之中。一方面，将既有的物质遗存加以保留再利用，保留了场地上的工业遗存，并挖掘展示了各厂区原有的历史故事，开辟成为了拥有工业文明记忆且与城市生活密切相连的滨水公共空间，如渔人码头防汛闸门、老码头地面肌理、钢质栓船桩、混凝土系缆墩等工业标志物；另一方面，以尊重场所技艺的方式进行有限度的介入，新建了钢廊架、钢栈道、钢廊桥、水管栏杆与灯柱等景观小品。

20th century, it had developed into one of the largest industrial bases in modern China and created numerous "the most" in the history of Chinese industry.

The urban renewal and redevelopment of Yangpu Waterfront will be divided into three sections: south, middle and north. The southern section starts from Qinhuangdao Road in the west, reaches Dinghai Road in the east, bordering Huangpu River shoreline in the south and Yangshupu Road in the north. The shoreline is about 5.5 kilometers long and the core waterfront area is about 1.8 square kilometers. In the past 20 years, the southern section of Yangpu Waterfront took the lead in experiencing the relocation of industrial storage, planning adjustment, land acquisition and divided construction of public space in different phases, and gradually transformed from industrial production shoreline to urban life shoreline. The exhibition site of 2019 Shanghai Urban Space Art Season is the public area that has been fully connected and open, and located between the first building level in the south section of Yangpu Waterfront and Huangpu River.

The industrial remainings of Yangpu Waterfront are large in scale and concentrated in distribution, with high historical and cultural value, scientific and technological value and artistic aesthetic value. In the south section of Yangpu Waterfront, the planning suggests to preserve a total of 66 historical buildings in 24 places, with a total construction area of 262,000 square meters. In addition, it also preserves a large number of distinctive industrial relic facilities and equipment. Many of these industrial remainings were very important buildings in the industrial history of China, such as:

The largest thermal power plant in the Far East — Yangtszepoo Power Plant (1913);

The earliest waterworks in China —Yangtszepoo Waterworks (1883);

The longest steel structure dock factory in modern times — Yangtszepoo Factory of Anderson, Meyer & Co. (1921);

The earliest factory building with reinforced concrete structure in China — the saw-roofed spinning wheel workshop of Jardine, Metheson & Co. Cotton Mill (1911);

The earliest multi-storey factory building with steel structure in China — No.1 Boiler Plant of Riverside Power Station (1913);

The highest modern steel-frame structure workshop in modern times — No.5 Boiler Plant of Riverside Power Station (1938);

And so on.

The continuous industrial zone along the waterfront used to be the "separation wall" between Huangpu River and the urban living space. The construction opens up the closed production space and turns it into an open living space, completely changing the state of urban space that we can not see the river when facing it and actually realizing the purpose of returning the river to the people. At the same time, the landscape design of connection project is committed to excavating the cultural background and place memory of the industrial heritage, seeking to reintegrate the original characteristic spaces and place characteristics of the industrial area into the urban life. On one hand, the existing material remainings are preserved and reused on the basis of preserving the industrial relic on the site, with original historical stories of each area are excavated and displayed, these remainings are developed into waterfront public spaces with the memory of industrial civilization and closely connected with urban life, such as: flood gate at Fisherman's Wharf, paving texture of old wharves, steel bolted boat piles, concrete mooring cleats and other industrial symbols; on the other hand, limited intervention is carried out with respect to the place,like steel-structure corridor, steel trestle, steel corridor bridge, water pipe railing, lamp post and other landscape features are built.

主展馆片区——毛麻仓库、小白楼、船厂船坞及周边地区

主展馆片区位于杨浦滨江南段的起始区段，聚集了一系列曾经享誉沪上的工业遗存，包括 1882 年建成的天章造纸厂、1896 年建成的英商怡和纱厂旧址、1900 年建成的上海船厂、上海毛麻纺织联合公司等等。主展馆片区由毛麻仓库，以及其北侧的小白楼、其东侧的船厂的两个船坞及周边地区围合而成。

其中毛麻仓库，以及其北侧的小白楼为主展馆。毛麻仓库位于杨树浦路 468 号，临江而建，于 1920 年由公和洋行设计。其立面工业特征明显，是杨树浦路上中国民族工业的印记。毛麻仓库所在的厂区最初在 1898 年由一个德国商人创办，叫"瑞记纱厂"，1918 年被英国人接管，更名为"东方纱厂"。1928 年民族资本兴起，上海当时的荣氏家族收购了东方纱厂，更名为"申新七厂"，见证了申新七厂这一中国最大的民族资本企业由盛转衰的历史。1949 年新中国成立以后，申新七厂成为国营厂，1951 年更名为"上海第二十棉纺织厂"，经过纺织系统内部调整，归到上海丝绸工业公司，1959 年定名为"国营上海第一丝织厂"。在 2002—2003 年，上海的纺织系统产业改造升级，整个厂区并入上海船厂范围内。毛麻仓库是滨江带现存建筑中最早的无梁楼盖仓库。小白楼紧邻毛麻仓库，原为上海船厂厂房，建于解放初期，本次作为展期的工作楼使用。

上海船厂船坞是上海历史最悠久的船坞之一，由德资的瑞镕船厂于 1900 年开挖，专造浅水船、拖船、驳船和游览船；1936 年同祥生船厂、耶松船厂等合并成为英联船厂，成为当时中国拥有最多船坞的船厂；1954 年并入上海船舶修造厂；1985 年再次改名为"上海船厂"。船厂内留存有双船坞，其中大船坞长 260 米，宽 44 米，深 11 米；小船坞长 200 米，宽 36 米，深 10 米。这两座船坞及其周边的塔吊群是该区域标志性的亮点，拥有工业区独特的场所感和艺术震撼力。2007 年 11 月 6 日，中国第三代极地科学考察破冰船"雪龙号"升级改造后交船离厂。而船厂船坞作为艺术展示场所全新亮相本次艺术季。

Main Exhibition Area – Maoma Warehouse, Xiaobai Building (Office Building), Shipyard Dock and surrounding area

Located at the beginning of the south section of Yangpu Waterfront, the main exhibition area contains a series of industrial relics that were once famous in Shanghai, including Tianzhang Paper Mill built in 1882, the former site of Jardine, Metheson & Co. Cotton Mill built in 1896, Shanghai Shipyard Co., Ltd. built in 1900, Shanghai Wool & Hemp Textile Union Company and so on. The main exhibition area is enclosed by Maoma Warehouse, Xiaobai Building to the north, two docks of the shipyard to the east and surrounding area.

The main exhibition is held in Maoma Warehouse and Xiaobai Building on its north side. Located at No.468 Yangshupu Road, next to the riverside, Maoma Warehouse was designed by Palmer & Turner in 1920. Its facade has obvious industrial characteristics, which is the mark of Chinese National Industry on Yangshupu Road. The factory where Maoma Warehouse is located was originally founded by a German merchant in 1898, called "Arnhold Karberg & Co. Cotton Mill", and was taken over by the British in 1918, renamed "Dongfang Cotton Mill". In 1928, with the rise of national capital, the Rong family in Shanghai purchased Dongfang Cotton Mile and renamed it as "Sung Sing Cotton Mill No.7 Plant", which witnessed how Sung Sing Cotton Mill No.7 Plant, the largest national capital enterprise in China, develop from prosperity to decline. After the establishment of the People's Republic of China in 1949, Sung Sing Cotton Mill No.7 Plant became a state-run factory and was renamed as "Twentieth Shanghai Cotton Plant" in 1951; due to internal adjustment of the textile system, it was subordinated to Shanghai Silk Industry Company, and in 1959 it was officially named with "State-run First Shanghai Silk Mill". In 2002 to 2003, with the transformation and upgrade of the textile industry in Shanghai, the whole plant was incorporated into Shanghai Shipyard Co., Ltd. Maoma Warehouse is the earliest storehouse with flat slab among the existing buildings along the waterfront. Xiaobai Building adjacent to Maoma Warehouse used to be the factory building of Shanghai Shipyard Co., Ltd., built in the early days of liberation, and this time it is used as the office building during exhibition.

The dock of Shanghai Shipyard Co., Ltd. is one of the oldest docks in Shanghai, which was excavated by German-owned New Engineering and Shipbuilding Works Ltd. in 1900, specializing in shallow water boats, tugboats, barges and sightseeing boats. In 1936, it merged with Xiangsheng Shipyard and Farnham & Co., S. C., and became the United Kingdom Shipyard with the biggest number of docks in China at that time. In 1954, it was incorporated into Shanghai Ship Repair Factory. In 1985, it was renamed as "Shanghai Shipyard" again. There are two docks remained in the shipyard, of which the big one is 260 meters long, 44 meters wide and 11 meters deep, while the small one is 200 meters long, 36 meters wide and 10 meters deep. The two docks and surrounding tower cranes are iconic highlights of this area, with a unique sense of place and artistic shock of the industrial zone. On 6 November 2007, China's third generation of polar research icebreaker "Xuelong" was handed over after being upgraded and renovated. The shipyard docks, as a place for art exhibition, made a brand-new appearance in this art season.

以编织的方式保持原真性——毛麻 仓库建筑的改造

Keep the Authenticity by weaving — Renovation of Maoma Warehouse Buildings

刘毓劼

2019 上海城市空间艺术季主展馆毛麻仓库改造建筑师

Liu Yujie

Architect renovated Maoma Warehouse, the Main Exhibition Site of 2019 Shanghai Urban Space Art Season

毛麻仓库的百年历史

毛麻仓库的改造设计大概是从 2016 年的三四月开始的。当时上海 45 公里滨江贯通工程正在进行，杨浦区滨江段是一个非常重要的贯通段，毛麻仓库也在其中。

我们对毛麻仓库的历史做了大量的调查工作。我们首先收集了各个年代的上海历史地图，在 1927 年的历史地图上就出现了毛麻仓库。然后，通过大量文字、图片资料的收集，我们发现毛麻仓库所在的整个区域从杨树浦路起到江边曾经是一个完整的厂区。紧贴江边是仓库，所有的原材料从仓库运到厂房，然后加工为成品，再从黄浦江边运走，形成了一个完整的生产区域。

我们还在上海市城市建设档案馆找到了毛麻仓库的原始图纸。对整个项目来说，这是最宝贵的一份资料。这份图纸由公和洋行绘制于 1919－1920 年。公和洋行是上海当时一家著名的建筑事务所，外滩的历史建筑中有其很多代表作，比如汇丰银行、沙逊大厦。图纸标注表明，到 2020 年，毛麻仓库这座建筑正好建成一百年。在这一百年里，它经历了三个时期：从外国资本进入上海，到民族资本和民族工业的兴起，再到国营建设时期的高峰。

建筑的改造与再利用

毛麻仓库始建于 20 世纪 20 年代，最大的特点是体现了当时的生产需求和技术发展特征。

首先，建筑的造型相对单一，平面形状规则。滨江一侧是简洁的立面，因为仓库比较封闭，立面上也没有太多的外窗和装饰。靠近厂区的一侧有一个宽大、平缓的楼梯，满足当时货物在上下运输时靠工人徒手搬运的功能需求。

其次，建筑结构很有特点，是当时典型的板柱结构。简单来说，这种结构和现在的梁柱体系不同，是柱子通过一个扩大的柱帽支撑楼板，从外墙上就非常清晰地反映出这种结构体系。板柱

Century-old Maoma Warehouse

The renovation design of Maoma Warehouse probably started in March or April of 2016. At that time, Shanghai 45 km Waterfront Connection Project was under construction, while the section in Yangpu District was a very important part, where Maoma Warehouse is located.

We did a lot of research into the history of Maoma Warehouse, starting from collecting historical maps of Shanghai in various eras, and found Maoma Warehouse on the historical map of 1927. Then, through the collection of a large number of texts and pictures, we found that the whole area where it is located now, from Yangshupu Road to the riverside, used to be a complete factory. The warehouse is close to the river, from which all the raw materials were transported to the workshop, processed into finished products, and then transported from the bank of Huangpu River, forming a complete production area.

We also found the most valuable piece of information for the whole project in Shanghai Urban Construction Archives, the original drawings of Maoma Warehouse. This drawing was made by Palmer & Turner between 1919 and 1920, which was a famous architecture firm in Shanghai at that time with many representative works amoung the historical buildings on the Bund, such as the building of HSBC and Sassoon Building. Indicated in the drawing that by 2020, Maoma Warehouse will be exactly 100 years old, after experiencing three periods: from foreign capital coming into Shanghai, to the rise of national capital and national industry, and to the peak of state-run construction period.

Renovation and Reuse of the Building

Maoma Warehouse was first built in the 1920s, greatly reflecting the production demand and technical development characteristics at that time.

First of all, the shape of this warehouse is relatively simple with regular floor plan. The afacde facing the river is simple and quite closed without too many exterior windows nor decoration. There is a wide and gentle stair near the factory area, for workers to transport goods with barehands at that time.

Secondly, the typical slab-column structure at that time is used in this building. This structure differs from the current system of beams and columns, in that the columns support the floor with an expanded column cap, which can be clearly reflected from exterior walls. Filler walls between the slabs and columns are made of red brick. We made a careful screening and analysis of the external wall at the site, stripping off the whitewashes and repairs in the 1970s and 1980s to reveal the original red bricks. The additions inside were also removed, except one external slide, which may have been added later for easier delivery, and the elevator were preserved. In this process, our working principle is to "repair the old as original to preserve the authenticity", with the aim of maintaining the authenticity of the building to the greatest extent.

之间的填充墙采用的是红砖。我们在现场对外墙做了仔细的甄别和分析，把七八十年代的外墙粉刷和修补处剥离，露出原来的红砖。还将室内加建全部拆除，仅对外部的一条滑道——可能是后来为了运货方便而加建的，和后来加建的电梯进行了保留。在这个过程中，我们的工作原则叫"修旧如故，以存其真"，目的是最大程度地保持建筑的原真性。

毛麻仓库原本只是一个工业建筑，没有特别考虑到人的使用或建筑的外观。但是从现在滨江绿化空间的完整性来看，需要让更多的公众去使用这座建筑，将更多新的时尚功能植入进去，以满足新的使用要求。因此在这一点上我们考虑的是：怎样把毛麻仓库的建筑本身以及周边场地的多种元素重新编织起来，以提升使用价值。巧妙的是"编织"这个概念也正好契合毛麻仓库所处的纺织工厂的历史特质。

我们从历年航拍照片分析毛麻仓库周边厂区的建筑建造与分布情况，发现仓库北侧至今仍被保留的一间厂房建于 1979 年。这间厂房反映了 20 世纪 70 年代工业建筑的典型结构特征——排架结构，而且它的预制构件非常精巧。因此，毛麻仓库及北侧车间这组建筑跨越了两个有代表性的建造时期：一个是 20 世纪 20 年代的建筑——板柱结构、红砖外墙；一个是 20 世纪 70 年代的建筑——排架结构、水泥拉毛外墙。这两个房子相互并置，相互呼应，相互印证，便成为一个微缩的滨江历史片段，反映出不同的历史时期的特征。于是，我们尝试着把这些元素都编织到建筑的改造中，让历史的多元性更加生动地呈现出来。

毛麻仓库东侧的两个船坞是 2019 年城市空间艺术季的主展场之一；北侧的绿地还在规划中；西侧是已经成型的滨江公共区域；南侧是上海的名片——黄浦江。我们把毛麻仓库二层沿江的墙体打开，使人在这里能够和黄浦江产生互动，在毛麻仓库的东侧和西侧增加了坡道，使其像丝绸一样正好通过毛麻仓库的二层，把整个滨江东西两侧带有高差的场地有机地联系在一起，形成连续的公共开放空间，为新功能的植入打下基础。对于遗留下来的北侧厂房，通过将其底层局部打开，使船坞到北侧的绿地、西侧的公共区域变成一个连续的人行流线空间，把人的活动充分纳入进来。这种把原来的工业建筑融到新的功能当中，去和周边的场地产生互动的手法，也是"编织"这个概念所希望实现的。

毛麻仓库作为仓储建筑，柱网非常规则，能承受的荷载较大，具备举办展览等大型公共活动的先天条件。为了配合展览，主要从三个方面对结构和设备进行了优化：首先，在不影响保护原则和现有空间的前提下，在结构上采取了必要的加固措施；其次，将现代化的水、电、空调设施植入进去，并尽量高效集中，留出完整的使用空间；最后，考虑到展览期间以及今后的使用，对原有的电梯设施进行改造，展览开始以后，又结合艺术家的要求做了相应调整。

Maoma Warehouse as an original industrial building, was not deisgned for the public to use nor appreciate its appearance. However, considering the integrity of the green space along the waterfront, it is necessary to allow more public to use the building and implant more new modern functions to meet the new requirements. At this point, we considered how to reweave the building itself and various surrounding elements to enhance its value. Coincidently, the concept of "weaving" also fits in with the historical nature of the textile factory where Maoma Warehouse is located.

Based on the analysis of the construction and distribution of buildings around Maoma Warehouse by aerial photos over the years, we found that a factory building preserved on the north was built in 1979. With typical structural feature of industrial buildings in the 1970s, it used bent frame structure and elaborate prefabrication. Thus, the Warehouse and the north side factory span two typical construction periods: one represnts the 1920s, with its slab-column structure and red brick exterior wall; the other one represents the 1970s, with bent frame structure and brushed concrete exterior. The two juxtaposed houses echo and confirm each other, which becomes a miniature of the waterfront history, reflecting the characteristics of different historical periods. Therefore, we try to weave these elements into the renovation of the warehouse, to vividly present the diversity of history.

The two docks on the east side of Maoma Warehouse are one of the main exhibitions of 2019 Urban Space Art Season; the green space on the north side is still under planning; the west side is well-performed waterfront public area; while the south is the name card of Shanghai, Huangpu River. We open up the wall on second floor of Maoma Warehouse facing the river, to let people interact with Huangpu River; a ramp is added at east and west side, silkly going through the secone floor, organically connecting the whole waterfront on both sides with a height difference, to form a continuous public open space as a good foundation for future new functions. As for the remaining northern part, the ground floor is partially opened up to make a continuous pedestrian circulation space from the dock to the green space on the north side and the public area on the west side, and people's activities are fully included. This way of integrating the original industrial building into the new function and interacting with the surroundings is also what the concept of "weaving" hopes to achieve.

Maoma Warehosue as a storage building with regular column system, can bear large load, which is also very suitable for holding large public activities like exhibitions. In order to cooperate with the exhibition, the structure and equipment were optimized from three aspects: first, the necessary structural reinforcement measures were taken without affecting the protection principle and the existing space; secondly, modern facilities for water, electricity and air conditioning should be implanted into it, and be concentrated as efficiently as possible, leaving a complete space for use; finally, the original elevator facilities should be upgraded considering the use during the exhibition and in the future, and corresponding adjustments were made according to the requirements of the artists after the exhibition began.

老建筑的价值判断与取舍

在对历史建筑进行改造的过程中，如何取舍，是一个长久以来最为纠结的问题。归根结底，是关于老建筑价值的判断。上海的建筑保护工作开展了30多年。我们在实践过程中进行了很多探索，其中一个主要原则是——保持建筑的原真性。珍贵的历史遗存应该保护、传承下去。但是单纯的保护实际上很被动，所以大家都在探讨怎么将其活化利用，怎么使其更好地为现代生活服务。这里就包含了价值取舍的问题，这是所有的矛盾、纠结、模糊和不确定的根源所在。不过随着整个社会对保护的认识在不断提高，各种制度也不断完善，我们的工作也变得越来越有意义。

城市空间与居民生活的关系

任何一个空间都会对人产生影响。如果一个空间对人的影响是积极的，人们就容易接受，反之就会比较抗拒。人们在一个喜欢的空间里，会更频繁地使用这个空间，也会有更多人来到这里，于是就产生了集聚。其实整个城市的发展就是各种各样的人集聚的过程，现在很多"网红空间"就是这样诞生的。因此，空间和人之间是一个互动的关系。

作为建筑师，可能更多是要了解、分析人对于空间的需求。在毛麻仓库的改造中，我们不仅分析原有建筑及其历史，还分析建筑周边的场地，及场地发展过程的历史，更和业主一起去设想今后各种可能的使用场景。建筑师在某种程度上进行的就是"编织"的工作，一种"求同存异"的系统架构工作。找到大家共同的需求，那么整个工作就能够顺利推进。

一座未来的"文化仓库"

在毛麻仓库改造项目中，对公众的开放和今后的再利用，是最重要的目标。目前对其未来的定位是文化仓库，希望能够通过保护和改造，让毛麻仓库和周边整个场地成为今后杨浦滨江的新亮点。

新亮点意味着，建筑的改造要体现四个方面的价值。一、历史价值。因为毛麻仓库是上海历史文化的沉淀，也是杨浦滨江百年工业发展的见证，应把历史还原给公众。二、原有的审美价值。因为这座仓储建筑代表20世纪20年代的工业与建造技术，代表了那个时代真实的生产情况，应唤起使用者和参观者的审美情绪。三、社会价值。毛麻仓库位于黄浦江畔，景色优美，作为再利用的探索，应使其更好地融入今后的社会生活，变成一个时尚和潮流集聚的地方。四、经济价值。作为滨江贯通工程的一个节点，毛麻仓库应在这片区域起到带动经济发展的作用。

如果浦东陆家嘴的建筑代表了上海的高度的话，那么毛麻仓库等杨浦滨江的工业遗存，代表了上海历史的厚度，呈现了另一个维度的上海。同时，毛麻仓库的改造不是一次单纯的建筑改

How to Judge and Choose the Value of an Old Building

In the process of renovating historical buildings, how to make a choice has been the most tangled problem for a long time. Ultimately, it's a judgment about the value of an old building. The action of protecting buildings in Shanghai has been going on for more than 30 years. We made a lot exploration through practice, and one of the main principles is to maintain the authenticity of the building. Precious historical relics should be preserved and passed on. However, pure protection is actually quite passive and hard to serve for modern life, so everyone is discussing how to activate and utilize it. This involves the question of how to make choice, which is the root of all contradictions, entanglements, ambiguities and uncertainties. With continuous improvement of social awareness of protection and various systems, our work has become more and more meaningful.

The Relationship between Urban Space and Resident life

Any space can have an impact on people. People are more likely to accept spaces that have positive impact on them, otherwise they will resist. In a favorite space, people will use it more frequently, and more people will come to gather. In fact, development of the whole city is a process of gathering all kinds of people, and many "Internet Celebrity Spaces" nowadays are born in this way. Therefore, there is an interactive relationship between space and people.

As an architect, we may have to pay more attention to understand and analyze people's demand for space. In the renovation of Maoma Warehouse, we not only analyzed the original building and its history, but also the surroundings of the site, and the history of its development, and brainstorm with the client to imagine more possibilities of future scenarios. To some extent, the architect is engaged in the work of "weaving", a systematic architecture work of "seeking common ground while reserving differences". If we could find the common needs, the whole thing will move forward smoothly.

A Future "Cutural Warehouse"

In the renovation project of Maoma Warehouse, the most important goal is open to the public and future reuse. At present, it is positioned as a cultural warehouse in the future, and hopefully the whole surrounding area will become a new highlight of Yangpu Waterfornt through protection and renovation.

The new highlight means that the renovation of this building should reflect four aspects of value. First, historical value. The Maoma Warehouse is a page of Shanghai's history and culture, and also the witness of the century-old industrial development of Yangpu Waterfront, which should be restored to the public. Second, the original aesthetic value. This storage building represents the industrial and construction techniques of the 1920s, the actual production conditions of that era, which should evoke the aesthetic emotions of users and visitors. Third, social value. Located at the bank of Huangpu River with beautiful scenery, it should be explored and reused as a place where fashion and trends will gather, to better integrated into the future social life. Fourth, economic value. As a node of waterfront connection project, Maoma Warehouse should play a role in promoting economic development in this area.

If the buildings in Lujiazui in Pudong New Area represent the height of Shanghai, then the industrial relics along Yangpu Waterfront like Maoma Warehouse, represent the thickness of Shanghai's history and present a different dimension of this city. At the same time, the renovation of Maoma Warehouse is not just a simple architectural project, with the support by the activities of the urban space art season, it becomes more acceptable by the citizens, and more people begin to care about surrounding old buildings and understand the value of protecting historical buildings. This offers great help for further renewal and development of the whole area.

The theme of 2019 Art Season is "Encounter". In the area of Maoma Warehouse, it means encounter in space and time — a hundred-year-old history meets now; encounter of aesthetics — the technical beauty of industrial buildings meets modern art; and also encounter of places — waterfront urban space meet art space.

造，还有城市空间艺术季的活动加持，使这里可以更多地被市民接受，从而有更多人开始去关心周边的老建筑，去理解历史建筑保护的价值。这为整个区域的进一步更新和发展，提供了巨大的帮助。

2019 年的艺术季主题叫"相遇"。在毛麻仓库区域，相遇是一种时空上的相遇——百年以前的历史和现在的相遇，也是一种审美上的相遇——20 世纪 20 年代的工业建筑技术之美和当今的艺术之美相遇，还是一种场所的相遇——滨江城市空间和艺术空间的相遇。

项目信息	Project Data
项目名称：毛麻仓库改造项目	Project Name: Renovation of Maoma Warehouse
建筑设计：同济大学建筑设计院（集团）有限公司	Architecture Design: Tongji Architectural Design (Group) Co., Ltd.
主持建筑师：刘毓劼	Principal Architect: Liu Yujie
房屋建造年代：1920 年	Built Year: 1920
文物登记信息：参照上海市历史文化风貌区和优秀历史建筑保护条例进行保护	Cultural Relic Registration Information: Protection shall be carried out according to the Protection Regulations of Historic and Cultural District of Shanghai and Excellent Heritage Architecture
结构类型：4 层混凝土板柱结构（无梁楼盖）	Strcuture Type: 4 story of concrete slab-column structure (flat slab)
建筑面积：6600 平方米	Building Area: 6600m²
层数：4 层	Floors: 4
建筑高度：约 17.4 米（其中底层高 4.6 米，二、三层高 4.2 米，四层高 4.1 米）	Building Height: about 17.4m (ground floor 4.6m high, 2nd and 3rd floor 4.2m high, 4th floor 4.1m high)
改造时间：2016—2019 年	Renovation Time: 2016 – 2019

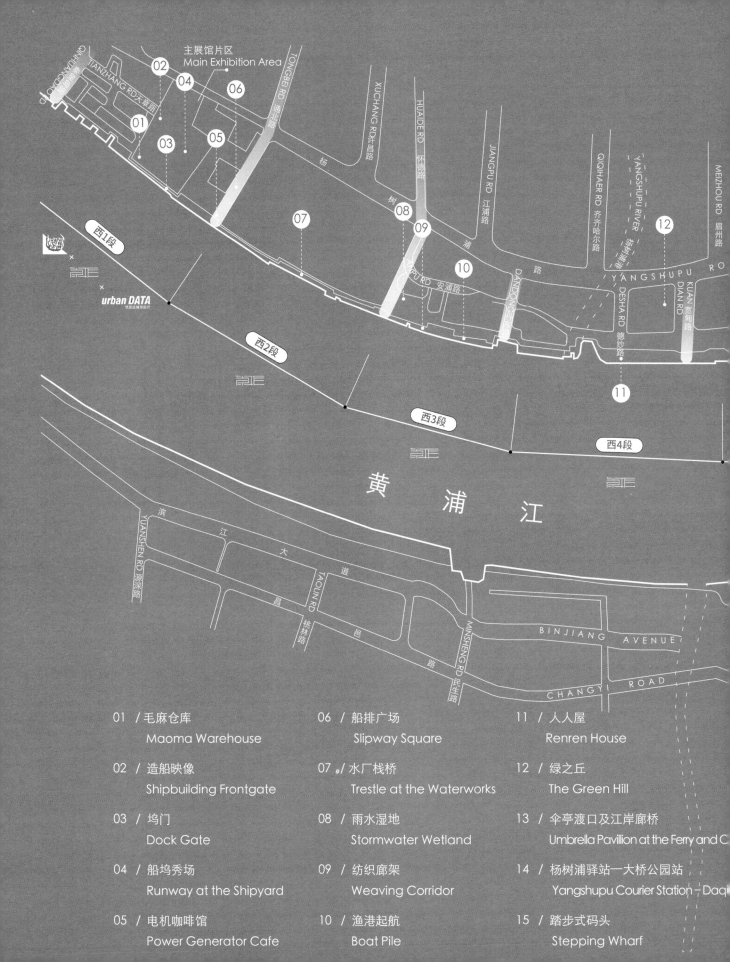

杨浦滨江公共
空间设计
Design of Yangpu
Waterfront Public
Space

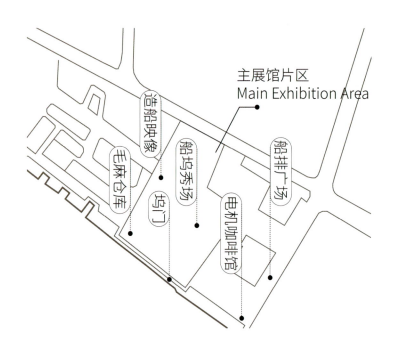

主展馆片区
Main Exhibition Area

造船映像
毛麻仓库
船坞秀场
坞门
电机咖啡馆
船排广场

西1段
（秦皇岛路—通北路）

1st West Section
(Qinhuangdao Road –
Tongbei Road)

W1

毛麻仓库　　Maoma Warehouse

本次空间艺术季的主展馆毛麻仓库是一栋拥有近百年历史的老建筑，于1920年由公和洋行设计，其钢筋混凝土无梁楼盖结构、简洁的红墙立面，凸显了上世纪20年代的技术特征和工业特色，是杨树浦路上中国民族工业的印记。

Maoma Warehouse, the main exhibition site of this space art season, is a historical building with one hundred-year-old history. It was designed by Palmer & Turner in 1920. Its reinforced concrete beamless floor structure and simple facade of red bricks highlight the technical and industrial characteristics of the 1920s, which is the mark of Chinese National Industry on Yangshupu Road

造船映像　　Shipbuilding Frontgate

设计师用脚手架搭建了入口大门。它具有一定的象征性，同时也解决了包括取票、志愿者服务、物品寄存等这样一些功能设施。这种脚手架的搭建方式能够快速建造，未来也能够快速拆除，是一种完全装配式的做法，既满足大型展览的需要，同时能够与船坞的大看台、钢木结构的报告厅，还有毛麻仓库、船坞，形成整个空间的序列，以一种艺术装置或者艺术品的方式呈现在观众面前。

The designer built the entrance gate with scaffolding, which is symbolic, but also provides enough spaces for ticket collection, volunteer services, storage and so on. This way of construction is both quick to build or remove, and also the practice of complete assembling, satisfying the need of large-scale exhibition. With the grandstand at the dock, lecture hall of steel and timber structure, and Maoma Warehouse, the dock, the scaffolding form the whole sequence of space, in the shape of art installation or artwork in front of the audience.

坞门　　Dock Gate

在坞门处，用无框拼缝的安全玻璃护栏替代了传统的金属护栏杆，同时保留了坞门平台上的牛角形，圆柱形系缆桩，并设置低矮细叶芒草等植被与之结合，巧妙地联系了通道和玻璃护栏。以枕木、碎石、耐候钢网格铺设路面，保证了安全畅通的同时，浦江对岸标志摩天楼组群，碧空之下的最美天际线，拉近到人们身边，一览无余。

At the gate, the traditional metal guardrail is replaced with the frameless patchwork safety glass guardrail, while the hornlike and cylindrical bollards on the platform are preserved, with plants like the short and thin-leaved Chinese silver grasses, cleverly connecting the channel and glass guardrail. The road is paved with sleepers, gravel and weathering steel grid to ensure safety. At the same time, the landmark group of skyscrapers on the other side of Pujiang River, the most beautiful skyline under the blue sky, are drawn close to audience and everything could be taken in one glance.

船坞秀场　　Runway at the Shipyard

船坞由德资的瑞镕船厂于1900年开挖，后改名上海船厂。2007年11月6日，中国第三代极地科学考察破冰船"雪龙号"升级改造后交船离厂。如今，这两座200多米长的船坞是该区域的标志性亮点，拥有工业区独特的场所感和艺术震撼力。在空间艺术季期间，船坞内呈现特色的装置和影像艺术作品，对公众开放体验。

The Shipyard was excavated by German-owned New Engineering and Shipbuilding Works Ltd. in 1900, which was later renamed as "Shanghai Shipyard Co., Ltd.". On 6 November 2007, China's third generation of polar research icebreaker "Xuelong" was handed over after being upgraded and renovated. The two docks with a length of 200 meters are iconic highlights of this area, with a unique sense of place and artistic shock of the industrial zone. During the art season, characteristic installations and video artworks presented in the shipyard docks are open to the public.

电机咖啡馆　　Power Generator Cafe

原为发电辅助用房，现已改造为杨树浦咖啡厅，人们手执醇香咖啡，围绕电机落座，一边看老式发电机，一边透过玻璃窗欣赏现代化的浦江美景。设计理念在尊重原有建筑二层高及面积的前提下，通过退台及屋顶平台塑造手法，使建筑赋予公共空间以不同层次，带给人们以多样的空间体验。

The former auxiliary room for power generation has now been renovated into Yangshupu Cafe. People holding mellow coffee, sitting around the motor, while looking at the old generator, they can enjoy the beautiful scenery of modern Huangpu River through the window. Respecting the height and area of the original second floor, the design concept creates different levels of public space and bring people a variety of space experience through the shaping techniques of the terrace and roof platform.

船排广场　　Slipway Square

杨浦滨江船厂段有一个特别的所在——船排遗址广场。这是一个让工业遗存活起来的奇妙景观，也是设计师们精心打造的公共开放空间。"扬帆起航，乘风破浪"，地面一座扬帆启航的船头高高扬起，好似奋力驶向远方；脚底玻璃之下，是一个曾经沉睡多年的船排。在这里，往来的游人既可以清楚地看到历史的遗痕，又可以感受昂首驶向未来的信心与勇气。

Slipway Reamins Square is a special place at Yangpu Waterfront shipyard section. It is a wonderful landscape that brings industrial remains to life, and a public open space that has been carefully created by the designers. "Set sail, ride the wind and waves", a ship on the ground with high-rised prow looks like ready for sailing; beneath the glasses under the feet is the slipway that has been sleeping for many years. Visitors can clearly see the traces of history, and feel the confidence and courage to head for the future.

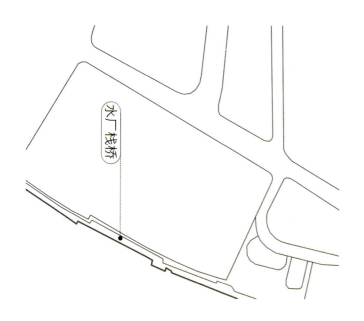

水厂栈桥　Trestle at the Waterworks

栈桥以"舟桥"为设计理念，构想一条船型结构的，与水厂历史风貌相协调的，悬浮于防撞柱子上的钢木结构的栈桥。在桥上行进的过程中仿佛在甲夹板上穿行，同时感受黄浦江景致与水厂历史风貌。桥面和栏杆扶手的一体化设计让木色连续完整，为黄浦江沿岸增添独特的元素，为杨浦滨江工业带增添温润的色彩。

The trestle uses "boat bridge" as the design concept, conceiving a boatlike structure in harmony with the historical features of the waterworks, which is a steel and timber structure trestle suspended on the anti-collision pillars. When walking on the bridge, you can have the same feeling walking on the deck, enjoying the scenery of Huangpu River and the historical scenery of the waterworks. The integrated design of the bridge deck and railings makes the wood colour continuous, adding unique elements to Huangpu River and warm atmosphere to Yangpu Waterfront Industrial Zone.

西3段

（怀德路—丹东路）

3rd West Section

(Huaide Road – Dandong Road)

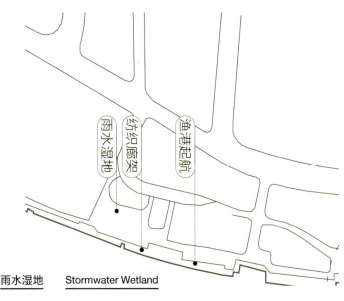

雨水湿地　　纺织廊架　　渔港起航

雨水湿地　　Stormwater Wetland

原先属于不同厂区的浮码头不可避免地存在高差和断裂，新建的漫步道则力图将其缝合。在原码头上种满芒草，形成凌空穿越草海的景观体验。透过底板局部透空的格栅网板能看到高桩码头粗壮的混凝土桩柱插入河床的状态，能观察到桥下黄浦江水的涨落变化，还能清晰地听到江水通过码头的夹缝拍打防汛墙的回响。

The new promenade tries to sew up the elevation differences and fractures that inevitably exist between floating piers of different sites. The original wharf is covered with miscanthus, forming a landscape experience of flying across the sea of grass. Through the partially open grid net underneath, we can see how did the thick concrete piles of piled wharf insert into the riverbed, observe the fluctuation of Huangpu River under the bridge, and clearly hear the echo of tides beating against the flood control wall through the crevice of wharf.

纺织廊架　　Weaving Corridor

廊架的建构原型源自原纺纱厂的整经机的工艺，将其重新演绎为座椅、攀爬索和遮阳棚等功能，纤细的钢柱和线性排列的钢索使廊架和坡道格外轻盈通透，与厚重的防汛墙和斑驳浮码头相比对，形成脱离于场地之上的漂浮态势。

The prototype of the corridor is derived from the process of warping machine in the original spinning mill, and is reinterpreted into programs such as seats, climbing cables and sunshade. The slender steel columns and linear steel cables make the corridor and ramp extremely light and airy, as if floating the site, compared with the heavy flood control wall and mottled floating pier.

渔港起航　　Boat Pile

大小不一的钢质栓船桩和老码头的粗糙肌理一同被保留下来。由于这些遗存物与新增围护栏杆存在位置上的冲突，因此每个墩座都需要针对性的节点设计，使栏杆有意避让栓船桩和系缆墩，使之成为滨江步道上时隐时现的景观小品。

Steel bolted boat piles of varying sizes are preserved along with the rough texture of the old wharf. Due to the conflict between these relics and the newly added railings in position, each pier seat is specifically designed to keep the railings away from the bolted boat piles and mooring cleats, making them a hidden landscape feature on the riverside promenade.

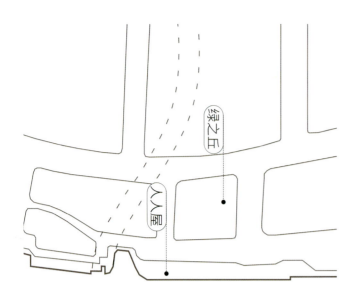

W4　W5

人人屋　Renren House

人人屋是杨浦滨江南段公共空间的一处滨水驿站。它是向每一位市民敞开的提供休憩驻留、日常服务、医疗救助的温暖小屋，故取名为人人屋。基于其所在区域是祥泰木行（始于1902年）的旧址，设计从一开始便确定了采用钢木结构的策略，希望温润的木质能够唤起人们对这段滨江历史的回顾与感知。

Renren House is a courier station in the public space of the south section of Yangpu Waterfront. It is a cosy house that open to every citizen to provide rest, daily services and medical assistance, hence be named as "Renren House". Considering the original site of China Import & Export Lumber Co., Ltd. (since 1902), the design strategy of using steel and timber structure was determined from the very beginning, hoping that the material characteristic of would arouse people's retrospect and perception of the history of this section.

绿之丘　The Green Hill

绿之丘是滨江复兴中对既有建筑烟草仓库实现的转型。设计时有条件地保留烟草仓库，对在单体建筑中垂直划分使用权属的新模式进行尝试，将其改造为集城市公共交通、公园绿地、公共服务于一身，被绿色植被覆盖、连通城市与江岸的建筑综合体所削切出来的生态之丘。

The Green Hill is the transformation of existing tobacco warehouse in the revitalization of the waterfront. The tobacco warehouse was conditionally preserved in the design, and the new mode of vertical division of use rights in one building is tried to transform it into an ecological mound cut out by the building complex covered with green vegetation, connecting the city and the waterfront, integrating urban public transportation, green park and public services.

西 5 段

（宽甸路—踏步式码头）

5th West Section

(Kuandian Road –
Stepping Wharf)

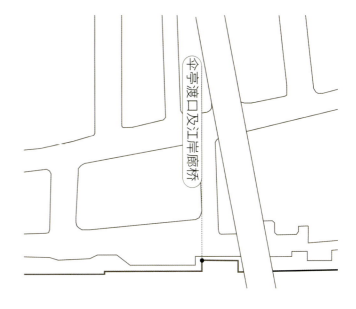

伞亭渡口　　Umbrella Pavillion at the Ferry

渡口设置伞状钢结构构筑物，加入顶棚，形成渡口空间。其间上升三个钢结构镂空伞状骨架，攀附绿植点缀。

The umbrella-shaped steel structure is set up at the ferry, with an added roof top to form the ferry space. Green plants climb up on the three steel-structure umbrella frames.

江岸廊桥　　Corridor Bridge by the River

通过钢构架形成衔接滨江漫步道的滨水廊桥。

Steel-structure waterfront corridor bridge is built to connect riverside promenade.

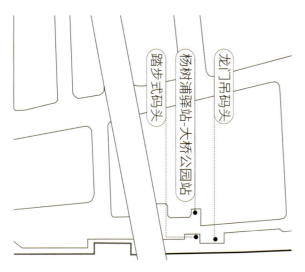

踏步式码头
杨树浦驿站-大桥公园站
龙门吊码头

东1段
（踏步式码头—广德路）

1st East Section (Stepping Wharf – Guangde Road)

杨树浦驿站—大桥公园站　Yangshupu Courier Station – Daqiao Park Station

原址为电站辅机厂西厂一处车间。将原车间厂房钢屋架整体保留使其成为内外通透的景观构架。构架内部利用原厂房局部墙体及楼板形成的空间，并置入一个 NLT 体系的木结构建筑，共同形成"屋中屋"。新植入的全木结构强化了西厂驿站的可识别性，半户外的屋架也为各种室外活动提供了适宜的场所。景观同建筑空间的并置利用得到充分体现。与构架一同保留的还有行车机身一部，工业电扇若干，更好地体现了历史遗迹。

The original site is a workshop in the West Plant of the Power Station Auxiliary Machinery Plant. The whole steel truss of the original roof is preserved as an airy landscape frame. The interior of the structure Takeing advantage of the space formed by the partial walls and floor slabs of the original workshop, a wooden structure building of NLT system is placed inside the frame, creating "a house within a house". The newly inserted wooden structure enhances the identification of the west courier station, and the semi-outdoor roof frame also provides a suitable place for various outdoor activities. The purpose of juxtaposing landscape and architectural space is fully reflected. Along with the frame, other historical relics are also preserved, like a bridge crane and a number of industrial electric fans.

踏步式码头　Stepping Wharf

据史料记载，踏步式码头为杨浦沿黄浦江最早的一批码头，采用水泥踏步式，与自然滩岸交错。景观完整保留此踏步式码头，并在其南面、农场局码头与电站辅机厂西厂码头之间新建栈道，并设置钢轨座凳，提供正面观赏踏步式码头的场所。

According to historical records, stepping wharfs are the earliest ones along Huangpu River in Yangpu, which use cement steppings intersect with the natural shore. The landscape deisgn preserved the intact stepping wharf and build a new trestle road at south between Farm Bureau Wharf and the West Plant Wharf of the Power Station Auxiliary Machinery Plant. Steel rail benches are set up to provide a place for admiring the stepping wharf from the front.

龙门吊码头　Gantry Crane Wharf

龙门吊码头位于电站辅机厂西厂旧址。该厂区设计通过低扰动介入的方式，设计了乌桕栈道，将现存的老码头连接起来，使水域步行系统连通。利用保留的工业设施如龙门吊、吊车等形成后工业码头景观。依托场地内现存的车间、老码头辅助用房等，将其改造为杨树浦驿站、休息凉亭等服务设施。在老码头辅助用房遗址上设计印记花园，再现老码头场地记忆。

Gantry Crane Wharf is located at the former site of the West Plant of the Power Station Auxiliary Machinery Plant. A low disturbance intervention approach is used in the design: a tallow wood trestle to connect existing old wharfs and the walking system by the water. The preserved industrial facilities, such as gantry cranes and cranes, are utilized to form a post-industrial dock landscape. Existing workshops and auxiliary rooms of the old wharf in the site, are transformed into Yangshupu Courier Station, rest pavilion and other service facilities. An imprint garden is designed on the remainings of the auxiliary room of the old wharf to reappear the site memory.

东2段

（广德路—水上桥亭）

2nd East Section

(Guangde Road –
Pavillion above the Water)

E1　E2

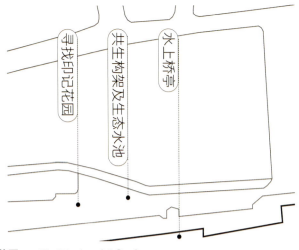

寻找印记花园　　Find the Imprint Garden

景区为国棉九厂沿江部分原址。该区通过历史考证，再现了厂区场地肌理和空间格局，保留了原厂现存的独具历史风貌的两座厂房，并选取原厂区内特征明显的仓库、车间等用房遗址，转化为富有当代体验氛围的主题性印记花园等景点。印记花园依照原厂房仓库遗址的轮廓，在新生场地中印刻而出，或临于江边，或嵌入绿地。原厂房外墙的印记统一用红色混凝土矮墙的形式再现，更强化这一概念。

The attraction is partly at the original site of Ninth National Cotton Plant along the river. Through historical research, this area reproduces the site texture and spatial pattern of the factory, preserves the two existing factories with unique historical features, and selects the original warehouse, workshop and other sites with obvious characteristics to transform into modern theme imprint garden and other tourist attractions. According to the outline of original factory and warehouse remainings, the Imprint Garden is engraved in the new site, either near the river or embedded in the green. The imprint of the exterior walls of the original building is reproduced in the form of red concrete parapets to reinforce this concept.

共生构架　　Symbiosis Architecture

共生构架原为锅炉厂（1953—1979年）期间建成的老厂房，因安浦路通过而需要被拆除。设计利用道路斜切角度，对原结构采取一半拆除、一半保留的加固策略，形成特殊形态和几何关系，结合新旧门窗洞口，打开屋顶，引入阳光和绿化，加入亲子沙坑、景观造坡，使墙里墙外形成一体，既保留了老锅炉厂的空间样貌，又转化成了供市民休憩活动的开放场所，成为建筑与景观、历史与城市的共生构架。

Symbiosis Architecture is used to be an old factory building built during the period of the boiler plant (1953–1979), which should be demolished for the passage of Anpu Road. Taking advantage of the oblique angle of road, the original structure is half dismantle and half kept to be reinforced, forming the special shape and geometry relations. Combined with the new and old windows and doors, the roof is opened up；bring in sunlight and green, adding parent-child sandpit and landscape slope, make the outside and inside the wall an organic whole. The space of old boiler plant is preserved and transformed into an open space for public recreation activities, and also a symbiotic frame between architecture and landscape, history and city.

水上桥亭　　Pavillion above the Water

水上桥亭位于国棉九厂与杨树浦煤气厂之间，将不同厂区的码头连成一气，形成了一条自西向东连绵不断的水上步道。混凝土栈桥每段长约42米，边缘微微外挑，做成了"甲板"的样子。

The Pavillion is located between Ninth National Cotton Plant and Yangshupu Gas Plant, connecting the docks in different factory areas, forming a continuous water path from west to east. Each section of the concrete trestle is about 42 meters long, with edges slightly protruding to create the appearance "deck".

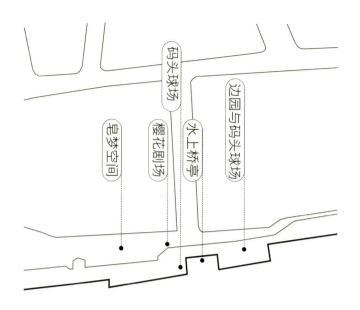

E3

皂梦空间　　Fantasy Bubble

皂梦空间位于景区内部的东侧，由原来的生产车间（中压水解楼）与环保科的污水净化池（包括调节池、格栅池、生物转盘池、气浮池、次氯酸钠池、观测楼）组成。环保科的原始围墙外饰面为水洗石，棕色、绿色与红色的玻璃与水洗石粘合在一起，具有强烈的年代感。远观，安静统一；近观，细腻的肌理混着年代的亲切感。水池的功能主题与肥皂结合，通过参与互动的方式，回顾历史与品牌，学习制皂工艺，将这组工业遗产转化为以新生活方式与浸入体验为核心的主题博物馆。

Fantasy Bubble is located at the east of the attraction, which is composed of the original production workshop (Medium Pressure Hydrolysis Building) and the sewage purification pool of the environmental protection department (including regulating pool, grille pool, biological rotary pool, floatation pool, sodium hypochlorite pool, observation building). The original wall of the Department of Environmental Protection is coated with pebble dash. Brown, green and red glasses mixed with the pebble dash, have a strong sense of age. See from a distance, quiet and unified; Close up, the delicate texture is mixed with the intimacy of age. The functional theme of the pool is combined with soap. Through participation and interaction, we can review the history and brand, learn the soap-making process, and transform this industrial heritage into a theme museum with new lifestyle and immersion experience as the core.

边园　　Riverside Passage

边园是对原煤气厂码头进行改造后的一处休憩与活动场所。它利用码头上现存的一堵90米长的混凝土长墙为基础，附着新建了跨越防汛墙和码头缝隙的坡道连桥、一个挑空的江景长廊、以及一处可以闲坐的亭。墙内是有些荒野感觉的小园林，墙外是在原有地面上磨出的一个旱冰场。它既充分保留了原有的工业印迹，又成功地将其转化为日常活动的场所。

Riverside Passage is a place for rest and activity after the renovation of original gasworks wharf. Based on an existing 90-meter-long concrete wall on the pier, across the gap between flood control wall and the pier, there is a newly built ramp bridge, a cantilevered corridor with river view, and a pavilion for sitting around. Inside the wall is a small garden with a sense of wilderness, while outside the wall is a roller-skating rink carved out of the original ground. Riverside Passage not only fully retains the original industrial imprint, but also successfully transforms it into a place for daily activities.

码头球场   Deck Sports Field

码头球场为于杨树浦煤场码头，原为杨树浦煤场的装卸码头，景观打造为休闲运动码头，设置沙滩排球场、码头篮球场、瞭望台等景点。球场周围的缆绳座凳，由天然麻质缆绳浸桐油后全手工缠绕而成，使景观细部更显码头特色。

The field is located at Yangshupu Coal Yard Wharf, which was originally the loading and unloading wharf of Yangshupu Coal Yard. The landscape is transformed into a recreational sports wharf, with beach volleyball field, dock basketball field, observation deck and other attractions. Rope benches around the field is made of natural linen rope soaked in tung oil and wound by hand, which makes the landscape details more characteristic of the wharf.

樱花剧场   Sakura Theatre

樱花剧场原为杨树浦装卸煤和物资的场地，景观结合原来煤堆的质感与剧场功能，并且联系整个滨江早春樱花带形成活泼的沿线露天聚会场所。下沉滑板乐园以自然形态的地形在樱花林中创造了一个仿若置身自然中的梦幻山谷，细节上材料结合水磨石的矿物肌理，与周边轻松自然的碎拼石材匹配后成为新与旧、自然与规整的完美融合，成为每个游园儿童记忆中充满工业运动魅力的梦幻处所。

Sakura Theatre is originally the place to load and unload coal and materials in Yangshupu. The landscape combines the texture of original coal piles with the function of theatre, and forms a lively open gathering place along the waterfront with the whole early spring sakura belt. Sinking skateboard park in the natural form of a terrain creates a natural dream valley in the forest. In detail, mineral textures of terrazzo match with the surrounding natural crushed stone and become the perfect fusion of old and new, natural and neat. This is the dream place full of industrial sport charm in the memory of every child.

纱泉广场及海绵走廊

东4段
（水上桥亭—腾越路）

4th East Section
(Pavillion above the Water – Tengyue Road)

E4 E5

纱泉广场　　Shaquan Square

纱泉广场位于整个基地的中部，是联系滨江景观带与开发地块商业建筑的主要通道。二级防汛墙与场地的高差通过一组大台阶消化，利用高差设计，石阶水池层层跌落，最终落于广场上形成景观浅水池。水池边模拟摇纱机设计了特色的互动景观装置，可控制喷泉的高度。水池内以纺织机零件为原型设计水车游乐装置，可为游人提供丰富的主题互动式游览体验。

Shaquan Square, located in the middle of the whole site, is the main channel connecting the waterfront landscape belt with the commercial buildings of the development plot. The height difference between the secondary flood control wall and the site is digested through a set of large steps, where the pool falls down layer by layer and finally lands on the square to form a shallow landscape pool. The simulated spinning machine at the pool features an interactive landscape installation that can control the height of the fountain. In the pool, a waterwheel amusement installation is designed based on the parts of textile machine, which can provide visitors with rich interactive experience of the theme.

海绵走廊　　Sponge Corridor

城市道路与防汛墙之间的绿地设计为雨水花园，减轻了风暴期间市政排水网络的负担，也使得硬混凝土墙不再是负面因素。此外，地埋式的雨水收集装置，为绿地提供了灌溉用水。林下的架空木栈道和休憩平台节点作为生态教育的空间，可以帮助人们进入并了解海绵城市的意义。

The green space between the city road and the flood control wall is designed as a rain garden, which relieves the burden on the municipal drainage network during storms and makes the hard concrete wall no longer a negative factor. In addition, a buried rainwater collection device provides irrigation water for the green space. The overhead boardwalk and resting platform under the forest serve as spaces for ecological education, which can help people enter and understand the meaning of sponge city.

东5段

（腾越路—内江路）

5th East Section
(Tengyue Road – Neijiang Road)

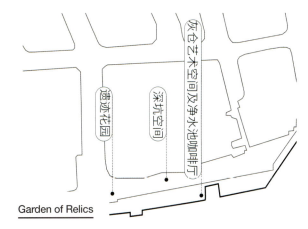

遗迹花园　　Garden of Relics

将原杨树浦电厂煤炭处理车间建筑群落的残柱及基础予以保留呈现。设计从钢板桩凹凸转折、相互咬合的工作机制出发，使其成为呈现历史、处理高差、组织流线、构造建筑、塑造景观的母题和线索。同时原建筑基础区域形成雨水自然下渗的湿地。

The remaining columns and foundations of the building groups of coal treatment workshop of former Yangshupu Power Plant are preserved and presented. The design starts from the working mechanism of concave-convex turning and mutual occlusal of steel sheet pile, making it become the motif and clue for presenting history, dealing with height difference, organizing circulation, constructing architecture and shaping landscape. At the same time, the foundation of original building area forms the natural wetland of rainwater infiltration.

深坑空间　　Deep Pit Space

保留电厂热力发电冷却水循泵取水深坑，形成别具特色的体验空间。

The deep pit for intaking and cooling water from the thermal power generation is preserved to form a unique space to experience.

净水池咖啡厅　　Purification Pool Cafe

建筑在原电厂污水处理清水池原址建设，原清水池基础予以保留并参与空间构成。新建结构完全同原有基础脱开，采用辐射状钢筋混凝土劈锥拱面壳体的结构形式，通过18瓣劈锥壳体连续相接形成环状整体。

The cafe was constructed on the original site of the sewage purification pool of the former power plant, and the foundation was preserved and participated in the spatial composition. The new structure is completely separated from the original foundation, adopting the structure form of radial reinforced concrete conoidal arch shell, through which the 18 sections of conoidal shell is continuously connected to form a whole ring.

灰仓艺术空间　　Ash Gallery

由原来杨树浦电厂的干灰储灰罐改造而成，通过架设新的交通体系，注入观光展示功能，创造出漫游登高远眺的滨江活动综合体。三个黄灰色相间、巨大而醒目的干灰储灰罐位于码头最东端，是电厂段的重要工业遗存。在拆除外围护结构后，对原有下部混凝土结构和上部钢结构分别加固，在中部形成一个平坦开阔的功能间层，为日后的功能拓展留下伏笔。三个灰罐中的两个经过重新划分层面后作为展示空间，另一个则以盘旋而上的坡道兼具交通和展示功能。罐体的外部界面虽全部置换为新的构造，却依然保持黄灰色相间的金属立面特征。

Transformed from the original dry ash storage tank of Yangshupu Power Plant, by setting up a new traffic system and introducing sightseeing program, a waterfront activity complex is created for wandering, climbing and overlooking. Three large and striking yellow and gray dry ash storage tanks are located at the easternmost end of the wharf, which are important industrial relics of the power plant. After the removal of the outer enclosure, the original lower concrete structure and the upper steel structure were separately reinforced to form a flat and open functional interlayer in the middle, foreshadowing for the future functional expansion. Two of the three gray tanks are recomposed as exhibition spaces, while the other one serves both traffic and exhibition functions with a spiral ramp. Although the external interface of the tank is completely replaced with new structure, it still maintains the yellow and gray metal facade features.

2019 年 11 月 17 日下午，"遇见：百年滨江相遇当代设计"设计师专场以"快闪导览"作为开场，活动请到了参与杨浦滨江 10 段共 30 个点位的 40 多位设计师和志愿者们现场讲解设计背后的理念，在长达 5.5 公里的岸线与观众们不期而遇。

下半场的活动延续了主题分享与圆桌讨论模式，来自多家设计机构的主创们在东方渔人码头的"渔货廊架"先后发言，分享设计心得。这里也是城市空间艺术季公共艺术作品《方块宇宙》的所在地。当天的大师讲坛由章明、刘宇扬、刘毓劼、单文慧、杨晓青做主题分享，由高目建筑设计工作室主持建筑师张佳晶担任主持。

Encounter: Centruy-old Waterfront Encounters Contemporary Design

On the afternoon of 17 November 2019, designers'show of "Encounter: Century-old Waterfront Encounters Contemporary Design" used "Flash Mobs Tour" as the opening, inviting more than 40 designers and volunteers who has participated in activities of 10 sections of Yangpu Waterfront, a total of 30 sites to present their design idea, encountering with the audience at 5.5 km long coastline.

Second half of the event continued the mode of theme sharing and round-table discussion, principals from many design institutions has made a speech to share design experience at "Fisherman's Corridor" in the Oriental Fisherman's Wharf. This is also the site of the public art work *Cube Universe* of the Urban Space Art Season. On that day, the master forum was hosted by principal architect Zhang Jiajing from Atelier GOM and shared by Zhang Ming, Liu Yuyang, Liu Yujie, Shan Wenhui and Yang Xiaoqing.

章明

杨浦滨江南段公共空间贯通工程总设计师；
2019 上海城市空间艺术季总建筑师；
同济大学建筑与城市规划学院建筑系副主任、教授、博士生导师；
同济大学建筑设计研究院（集团）有限公司原作设计工作室主持
建筑师

Zhang Ming

Chief Designer of South Section of Yangpu Waterfront Public
Space Connection Project; Chief Architect of 2019 Shanghai
Urban Space Art Season;
Vice Director, Professor, PhD Supervisor of Department of
Architecture, College of Architecture and Urban Planning,
Tongji University
Chief Architect of TJAD Original Design Studio

"上海 2035 总体规划"中提出上海要打造卓越的全球城市，突出中央活动区的全球城市核心功能。黄浦江两岸滨水空间作为中央活动区的重要组成部分，将把封闭的生产岸线转变成开放共享的生活岸线，这将对上海的城市空间格局产生重大影响。在参与杨浦滨江的建设时，经过对杨浦滨江从历史到现存状态的了解，我们制定了一个目标：要打造 5.5 公里、连续不间断的工业博览带，以工业为核心，打造一个生活化、生态型的、智慧型的、具有国际水准的公共开放空间岸线。

另一个重要的概念是"锚固于场地"的更新模式。在设计过程中强调场所精神，强调既有环境的物质留存，要做有时间厚度的杨浦滨江公共空间景观。人当然是场所的主题，但更重要的是场所的诗意呈现，支撑这种理解的核心是有限介入与低冲击开发。因为滨江资源已经很好，如何对待它才是关键。所以我们的一些设计元素，包括杨浦滨江很有特色的水管灯、栏杆、雨水花园上的廊架、纺车廊架等，用了很多工业上常用的要素，重新进行组织，按照我们设计师的理解和本身对美学的追求，重新解构之后共同营造出杨浦滨江的特质和风貌。

2019 年上海城市空间艺术季选用船坞地区作为主展馆片区，开幕式放在船坞当中举办。场地中的小船坞作为开幕式的场所，有艺术家的作品在里面呈现，还有大型的多媒体装置艺术。而大船坞我们希望它能够在空间艺术季期间，甚至艺术季之后举办各种各样的船坞论坛，包括承接一些类似于服装走秀等这样的时尚活动，真正把原来修船、造船的这样一个空间，转化为当下能够使用的时尚的空间。

如何看待这个区域的过去，可能就映射了这个区域未来会向何处去。时间在不同的历史阶段留下了痕迹，我们要做的工作就是如何把蒙在这个历史信息上的那些浮尘掸去，让历史的信息能够呈现出来。我觉得这就是我们当下的建筑师对历史的一种认知，让时间的厚度慢慢地在城市当中呈现出来。

In "Shanghai 2035 Plan", it is proposed to make Shanghai an outstanding global city highlighting the core functions of the central activity area. As an important part of the central activity area, waterfront space on both sides of Huangpu River will transform the closed production shoreline into an open and shared living shoreline, which will have a significant impact on the urban spatial pattern of Shanghai. When involved in the construction of Yangpu Waterfront, through understanding the history and current condition, we set a goal to build a 5.5-kilometer-long, continuous industry expo belt, centred with industry, creating a life, ecological, intelligent public open space shoreline of international standards.

Another important concept is the renewal model of "anchored to the site". In the design process, the spirit of the place and the material preservation of existing environment should be emphasized, so as to make the public landscape of Yangpu Waterfront with time thickness. People are the theme of the place, while the poetic presentation of the place is more important，which is mainly supported by limited intervention and low impact development. Having great waterfront resources, the key is how to treat with it. Some elements of our design, including the distinct pipe lights of Yangpu Waterfront, railings, the corridor above the rain garden, spinning wheel corridor and etc., used a lot of common industial elements, according to our designer's understanding and pursuit of aesthetics, which are reorganized and deconstructed to jointly create the characteristic and style of Yangpu Waterfront.

In 2019 Shanghai Urban Space Art Season, the Shipyard area was selected as the main exhibition area, and the opening ceremony was held in the dock. The small dock in the site was served as the venue for the opening ceremony, where artists' works were presented, as well as large-scale multimedia art installation. While the big dock was hoped to hold a variety of Dock Forums during or even after the Art Season, including undertaking some fashion activities like fashion show, so as to truly transform the original space of ship repair and shipbuilding into a fashion space that can be used nowadays.

The way we treat the past of this region may be a reflection of where it will go in the future. Time has left traces in different historical stages and what we need to do is to dust off the historical information so as to be presented. I think this is current architects' perceptions of history to let the thickness of time slowly presented in the city.

张斌

致正建筑工作室 Atelier Z+ 创始人、主持建筑师

Zhang Bin

Founder and Principal Architect of Atelier Z+

东 1 段—龙门吊码头
1st East Section – Gantry Crane Wharf

东 3 段—皂梦空间
3rd East Section – Fantasy Bubble

致正工作室负责杨树浦六厂滨江公共空间更新总控、上海电站辅机厂西厂、国棉九厂和上海制皂厂以及所有的水上部分。杨树浦工业带诞生了一大批中国最早的基础设施行业，比如说水厂、煤气厂、发电厂。它支撑了上海的发展，以及整个中心城区的繁荣。我们见证了工业带背后整个杨树浦巨大的功能集聚区，一直延续到现在。现在用滨江带去率先启动再生，这对建筑师来说，是一个巨大的机会，也是非常大的责任。

从杨浦大桥衔接出来，穿过码头栈桥，到国棉九厂区域，有一个杨浦滨江现存最老的码头。老码头的岸上，我们留了一个工业架构，原来是一个小厂房。我们觉得小厂房在沿江第一线，同时是从大桥的路域转到水域区的一个关键点上，所以留了一个钢结构的架构。这个架构里面还有桁车，上面做了一部分的阳光板覆盖，下面做了休闲空间，里面还容纳了为地区服务的设施。

国棉九厂的老厂房可能是目前作为优秀历史建筑保存下来的唯一一个大型的全钢结构早期厂房，而且是两层楼。景观是它的陪衬，是它的前场，所以我们设定了一个事件型的广场空间。

对于制皂厂，我们保留了一堆污水处理工艺水池，一个中压水解楼，沿江一个五层楼高的监测楼。建筑下部通过钢管通道将不同的水池串联互通，形成一系列的明暗转换、高低错落、内外翻转的空间。地下空间的紧凑感与登高望江的开阔感形成强烈的对比，由此创造出地上与地下两条不同的浏览体验路径。

Atelier Z+ is responsible for the general control of waterfront public space renewal of the Sixth Yangshupu Factory, the West Plant of Shanghai Power Station Auxiliary Machinery Factory, Ninth National Cotton Plant and Shanghai Soap Factory, as well as all the parts above water. The Yangshupu Industrial Belt is the home to many earliest infrastructure industries in China, such as waterworks, gas plants and power plants. It has underpinned the development of Shanghai and the prosperity of the entire downtown area. We witnessed the huge functional agglomeration of Yangshupu behind the industrial belt, which continues till today. To the architect, it is a great opportunity as well as huge responsibility to take the lead in starting regeneration with the waterfront.

There located the oldest wharf of Yangpu Waterfront, from Yangpu Bridge, through the pier trestle, to the area of Ninth National Cotton Plant. On the bank of the old wharf, we preserved an industrial structure, which used to be a small workshop. We found it located at the first line of waterfront and a key point where the road of the bridge transferred to the water area, so a steel structure was kept. This structure also contains a hoist, which is partially covered with polycarbonated panels, and a leisure space is created below, also housing facilities serving the area.

The old factory building of Ninth National Cotton Plant is probably the only earliest large-scale, all-steel structure factory that has been preserved as an excellent heritage architecture, and it has two stories., Therefore, we set up an event-based square space, making the landscape as its foil, its foreground.

As for the soap plant, we preserved a lot of wastewater treatment tanks, a Medium Pressure Hydrolysis Building, and a five-story monitoring building along the river. In the lower part of the building, different pools are connected through steel pipe channels, forming a series of spaces with light and shade conversion, high and low scattered, and inside and outside turning over. The compactness of the underground space is in sharp contrast to the openness of climbing up and looking over the river, creating two different experiencing paths for the above-ground and underground.

刘宇扬
刘宇扬建筑设计事务所主持建筑师

Pricipal Architect of Atelier Liu Yuyang Architects

东 2 段—共生构架
2nd East Section – Symbiosis Architecture

面对这个遍布工业历史遗存的场地，我们问自己"What to keep"，首先是保住场地都市丛林般的"野性"（Keep It Wild），其次是保住场地在地性和生态性中的"生鲜"（Keep It Raw），再次是保住"自然"（Keep It Natural），最后是保住"慢活"（Keep It Slow）。整个前期策划在定位上分为"水岸三段""城市四轴""杨树浦六厂（场地）"。一方面通过工业厂房尽可能地留下痕迹，另一方面利用景观的手法重新呈现这些厂区曾经严密的围墙。

我们把整个1.6公里分成若干个主题，电站辅机厂东厂这段就是以"市集"为一个主题，希望能形成周末创意市集。这个厂房曾是属于上海锅炉厂的一个老仓库，在最早的规划里是要被整体拆除的，因为它被一条规划道路切成了一半。我们参与具体设计的时候，做了一个折中的方案，允许一半的房子被拆，另外一半被保留，把留下的另一半当成一个景观构筑物处理，用钢结构的方式进行加固，并形成我们一个最核心的理念——共生构架。

当打开这栋老房子以后，我们发现它的范围是原来的一半，但视野是无限扩大的。我们后来有了一个非常好的想法，就是把外部的景观也带到原来的仓库空间里，使得内与外形成一种更模糊的状态，整个景观介入到建筑。因此，在"共生构架"场地旁，我们创造了一个纯景观的体验，现场地形原来就有一点微微下洼，我们借地形做了一个生态水池。这样一个自然的场景就能够同共生构架很有历史感、很有空间感的元素呼应，彼此邻近，又是截然不同的手法。

Facing this site full of industrial historical relics, we asked ourselves "What to keep". The first is to "Keep It Wild" like the urban jungle; the second is to "Keep It Raw" in the localization and ecology of the site; and then, "Keep It Natural"; finally, "Keep It Slow". The whole preliminary planning is divided into "Three Sections of the Waterfront" "Four Axes of the City" and "Sixth Yangshupu Plant (site)". On the one hand, we tried to leave as many traces as possible through industrial plants, and on the other hand, we used landscape strategies to re-present the once tight walls of these factories.

We divided the whole 1.6 kilometers into several themes, and the section of the East Plant of Power Satation Auxiliary Machinery Plant is themed as "Market", hoping to form a creative market on weekends. The factory, an old warehouse of Shanghai Boiler Plant, was originally intended to be demolished because it was cut in half by a planned road. When participating in the design, we made a compromised proposal that half of the building to be demolished and the other half to be preserved and treated as a landscape structure, reinforcing it with a steel structure, and forming one of our core concepts — Symbiosis Frame.

When the old building was opened up, we found it was half scale of it used to be, but the view was infinite. We then had a very good idea to bring the outside landscape into the original warehouse space, so that the whole landscape enters into the building, forming an ambiguous between inside and outside. Therefore, we created a pure landscape experience next to the "Symbiosis Frame" site, where the topography was originally slightly lowered and we used the topography to create an ecological pool. A natural scene like this is able to echo with the historical and spatial elements of Symbiosis Frame, which is close to each other but a completely different approach.

柳亦春

大舍建筑设计事务所主持建筑师、创始合伙人

Liu Yichun

Partner and Principal Architect of Atelier Deshaus

东 3 段—边园
3rd East Section – Riverside Passage

在杨浦滨江东段的沿岸景观设计里面，我们主要负责堆煤场和煤气厂这段，即杨浦滨江五期二段的最东段。"边园"是我们给堆煤场改造完之后所起的名字。我们想充分挖掘煤气厂的码头，作为一个工业场所的潜力。最吸引我的就是这堵墙，它是五期中最边远的一段，又是水岸边，当我看到这堵墙的时候，就觉得可以做成一个水边的园林。

这里原来是上海市煤气厂运煤的码头，承担的功能是轮船把煤从水路运过来，再堆到码头上。为了防止码头上的煤在装卸时散落到江水里，就在码头上建了一堵很长的混凝土墙。这堵墙总共有 90 米长，大概四米多高，前面的场地非常开阔。我们希望能够加强岸边跟码头的关系，特别想做一个连桥或者坡道的桥，可以跨越防汛墙伸到码头上，跟这长长的混凝土墙，构建出一个新的游览步行系统。长廊的下面，原来堆煤码头设计成了一个旱冰场。码头的溜冰场，墙上的长廊，都是直接利用堆煤场码头的构件。一个是地面，一个是墙面，这两个最基本的建筑构件成就了一个新的日常性生活场所—— 一个可以停留、可以经过的观景长廊。

在原来煤气厂的堆煤码头上，重新塑造一个新的日常生活空间场所，跟黄浦江建立一种新的关系，但这个新的关系仍然是渗透的，润物无声地融入到场所中。这是我比较喜欢的一种方式，一种可以让时间得以连续的方式。

In the landscape design of the east section of Yangpu Waterfront, we are mainly responsible for the part of coal yard and gas plant, namely the most eastern part of the second section of Yangpu Waterfront Fifth Phase. "Riverside Passage" is the name we gave to the coal yard after renovation. We want to dig into the full potential of the wharf of gas plant as an industrial site. What attracts me most is this wall, which is at the most remote section of Fifth Phase and also at the riverside. The time I saw this wall, I thought it could be made into a waterfront garden.

This used to be the wharf for Shanghai Gas Plant to transport coal, where ships carried coal from the waterway and then piled it. To prevent coal from falling into the river during loading and unloading, a very long concrete wall was built on the wharf. The wall is 90 meters long in total and about four meters high, with an open foreground. We hope to strengthen the relationship between the shore and the wharf, in particular by building a bridge or ramp bridge, which can extend over the flood wall to the wharf, and along with this long concrete wall, form a new sightseeing walking system. Under the long corridor, the former wharf for caol piling was designed into a roller-skating rink. The skating rink at the wharf, the long corridor along the wall, are all components that are directly utilized on the coal yard wharf. One is the ground, the other is the wall, these two most basic architectural components accomplished a new place for daily life —a sightseeing long corridor to stay or pass by.

On the wharf of the original gas plant, a new space for daily life is rebuilt to establish a new relationship with Huangpu River, but this new relationship is still permeable and slightly integrated into the site. This is a way that I prefer, a way that allows time to be continuous.

杨晓青

大观景观设计主持建筑师

Yang Xiaoqing

Principal Architect of Da Landscape

杨浦滨江的十二棉地块的特殊性在于它是杨浦滨江 5.5 公里岸线中第一块正式出让的地方，也完成了地块从工厂到新用途的转型。十二棉有 100 多年的历史，经历了几次转型。在 2018 年大观接手设计时，可以说地块里已经什么都没有，可能也是所有岸线里独一无二的、没有现状的基地。章明教授的 5.5 公里总体贯通方案对这一段设定了大致的方向，也给出了足够的留白。我们团队思考的是，过去地块作为棉纺厂将植物（棉花）转换为人们可使用的纺织物，现在我们是否可以反过来，用植物将工厂转换为人们可以使用的绿色空间，在某种程度上继续延续这块场地与人们生活的关联度。

场地中最显著的特征是代表原有工厂的斜线与代表沿江动线的横向两种机理的编织与融合，在这个序列中衍生出的种植形成了一段连续的绿色生态长廊，其中嵌入活动场地与休憩设施。中心的开放式广场中融入了可进入的互动水景，铺地和水景中用到的元素源于纺织机械上的零部件。考虑到儿童活动场地在滨江的稀缺，我们特地设置了针对不同年龄段使用的无动力游乐设备空间，并通过与地形、绿化的组合形成高参与度、趣味型的滨江活动体验。

这块区域没有太多历史保护建筑，这反而使得我们可以将重心放在人的体验和活动上，再通过多样化的形式表达，在离黄浦江最近的地方呈现出以绿色为主的舒适、可游玩的空间。

The particularity of the Twelfth Cotton Plot in Yangpu Waterfront lies in that it is the first piece of the 5.5 km coastline to be officially transferred, which also completes the transformation of the function from factory to new use. This plot has a history of more than 100 years and has undergone several transformations. When we took over the design in 2018, it could be said that there was nothing left on the plot, which was probably the only site with no current condition along the whole shoreline. Professor Zhang Ming's 5.5 km overall connection plan sets a general direction for this section and also gives enough blank space. Our team was thinking that in the past, the site was used as a cotton mill to convert plants (cotton) into fabrics that people can use, now whether we can do the reverse and use plants to convert the plant into green space for people, in a way to continue the relevance of the site to people's life.

The most particular feature of this site is the weaving and integration of the two mechanisms: the oblique line representing the original factory and the horizontal line representing the circulation along the river. The planting derived from this sequence forms a continuous green ecological corridor, which is embedded with activity areas and rest facilities. The central open plaza incorporates an accessible and interactive water feature, which uses elements from textile machinery in the paving and water feature. Considering the lack of activity spaces for children along the waterfront, we specially set up space for unpowered amusement equipment for different age groups, and through the combination of terrain and planting to create high-participation and interesting riverside activity experience.

The lack of historical preservation buildings in this area allows us to focus on human experiences and activities, which are expressed in a variety of forms, and present a green, comfortable, playable space adjacent to Huangpu River.

十二棉滨江景观俯瞰
Aerial View of the Twelfth Cotton Plot Waterfront

十二棉滨江中心开放广场
The Central Open Plaza of the Twelfth Cotton Plot Waterfront

刘毓劼

同济大学建筑设计研究院都境院副院长

Liu Yujie

Vice Head of TJAD Dujing Architectural Design Institute

毛麻仓库这个项目的设计大概是在 2016 年的三四月份开始的。当时上海要做滨江两岸的贯通，杨浦区滨江段是一个非常重要的贯通段，毛麻仓库也在其中，是整个杨浦滨江段的起点。这个面宽只有 54 米的仓库见证了完整的上海工业发展，从最早由德商创办，经历了民族工业时期，一直到解放后辉煌的国营时期。这座厂有自己的工业精神，能够代表上海的工业精神。毛麻仓库建于 20 世纪 20 年代，原来只是一个工业建筑，造型比较单一，空间也比较封闭，不太考虑到人的使用。但是放在现在的滨江，尤其在滨江整个绿化的空间当中，我们需要有更多的公众去使用它，有更多新的时尚功能植入进去，满足新的使用要求。

因此，我们考虑怎样能够把毛麻仓库建筑本身以及周边场地的情况特征，重新编织一下，提升它的使用价值。编织这个概念也是来源于场地和纺织厂的历史。我们把仓库沿江的一侧打开，既连通东西两岸，又连接从杨树浦路到杨浦滨江的线路。为了配合这次改造，我们在 20 年代的毛麻仓库和 70 年代的厂房之间重新架设了一个钢楼梯和平台。由于这个设计，两幢建筑能够更好地联系到一起。仓库屋顶上的木质平台是最后一笔，在那里能观赏到对岸浦东的景象。

西 1 段—毛麻仓库
1st West Section – Maoma Warehouse

The design project of Maoma Warehouse started around March or April 2016. At that time, Shanghai was pushing the connection project on two sides of the riverside. Yangpu Waterfront section was a very important part, where Maoma Warehouse is located, which was also the starting point of the whole Yangpu Waterfront section. With 54 meters width, this warehouse has witnessed the complete industrial development of Shanghai, from the earliest German merchant, through the period of national industry, to the glorious period of state ownership after the liberation. This factory has its own industrial spirit and can represent the industrial spirit of Shanghai. Maoma Warehouse was built in the 1920s, originally just an industrial building, with relatively simple shape and enclosed space, not much consideration of people's use. However, in present waterfront area, especially in the whole green space, we need more public to use it, and more new fashion functions implanted to meet new requirements of use.

Therefore, we thought about how to reweave the characteristics of Maoma Warehouse itself and the surroundings to enhance its value of use. The concept of weaving also comes from the history of the site and the textile factory. We opened up the side facing the river, connecting both the east and west banks and the road from Yangshupu Road to Yangpu Waterfront. To support this renovation, a new steel staircase and platform was inserted between Maoma Warehouse of the 1920s and factory building of the 1970s, making these two buildings more connected to each other. The final touch is a wooden platform on the roof, which offers a great view of Pudong across the river.

西 1 段—毛麻仓库
1st West Section – Maoma Warehouse

单文慧
优德达城市设计咨询有限公司规划设计总监

Shan Wenhui
Principal Architect and Planner of UrbanDATA

西 1 段—船排广场
1st West Section – Slipway Square

我们的设计重点在船厂区域，作为有规划背景的团队，更多地从城市空间、历史关系出发，形成设计策略，这是设计过程中最重要的部分。上海船厂浦西分厂的历史始于 1897 年，经过了几次变更，到 20 世纪 90 年代中叶，保持着技术上全国领先的地位。在设计的时候，压力就在于设计师应该留什么、创新什么，最终我们的愿景是通过公共空间的设计去诱发一个多样活动性场地，让使用者可以根据自己的需求创作活动场景，所以墙、铁轨、油罐这样的工业元素被保留了下来。

2017 年 1 月的时候，在施工现场挖掘出了一个百年历史的船排，于是我们团队赶紧修改设计，把船排保留下来，这也是城市更新项目会带来的惊喜。为此特地做了一个公共雕塑，寓意乘风破浪，也作为发现船排的纪念。其他的公共艺术雕塑也结合了工业元素处理，可以说，对整个滨江带的工业遗存都做了非常系统性的、结构性的保留。

Our design focused on the shipyard area. As a team with planning background, we formed a design strategy from the perspective of urban space and historical relationship, which was the most important part of this design process. Puxi Branch of Shanghai Shipyard Co., Ltd. began its history in 1897 and has undergone several transformations, but until the mid-1990s, it still maintained its leading position in technology in China. During the design process, the pressure on designer is what should be preserved and what should be innovated. In the end, our vision is to create a site for various activities through the design of public space, allowing users to create activity scenes according to their own needs, so industrial elements such as walls, rails and oil tanks are retained.

In January 2017, a century-old slipway was excavated at the construction site, and our team quickly revised the design to preserve it, which is also the surprise of the urban renewal project. A public sculpture was specially made, meaning to ride the wind and waves, and also as a memorial to the discovery of the slipway. Other public art sculptures are also combined with industrial elements, which means the industrial remains of the entire waterfront are preserved in a very systematic and structural way.

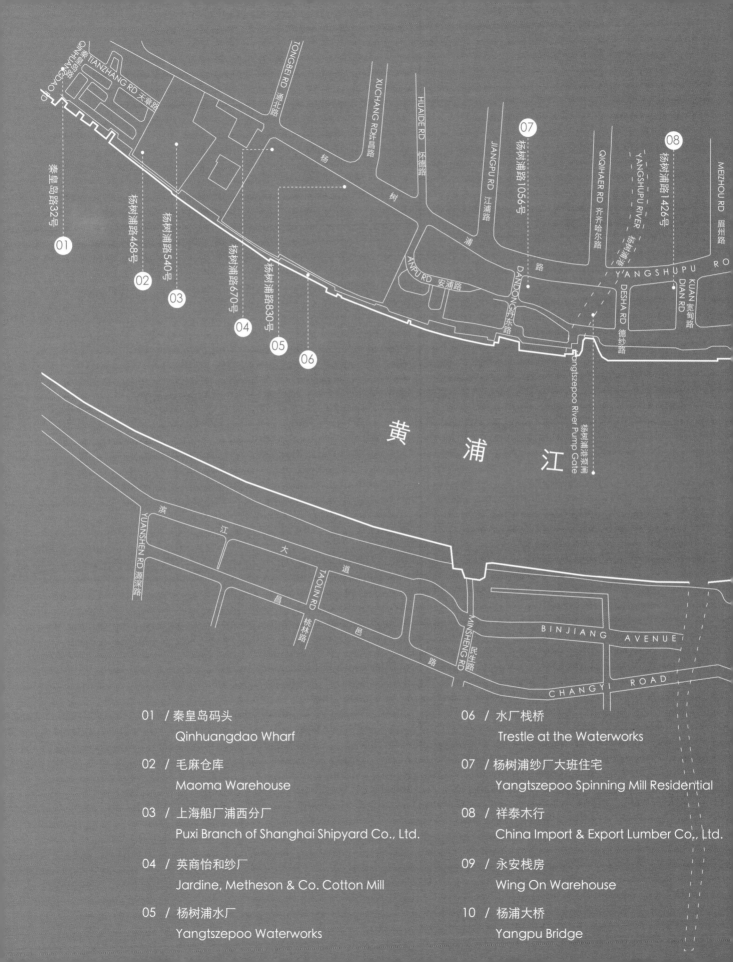

秦皇岛路32号
01

杨树浦路468号
02

杨树浦路540号
03

杨树浦路670号
04

杨树浦路830号
05
06

TIANZHANG RD 天章路

QINHUANGDAO RD 秦皇岛路

TONGBEI RD 通北路

XUCHANG RD 许昌路

HUAIDE RD 怀德路

杨 树 浦 路

JIANGPU RD 江浦路

杨树浦路1056号
07

ANPU RD 安浦路

DADONG RD 大东路

QIQIHAER RD 齐齐哈尔路

YANGSHUPU RIVER 杨树浦港

杨树浦路1426号
08

DESHA RD 德纱路

Yangtszepoo River Pump Gate 杨树浦港泵闸

KUAN DIAN RD 宽甸路

YANGSHUPU ROAD

MEIZHOU RD 眉州路

黄　浦　江

YUANSHEN RD 源深路

滨 江 大 道

TAOLIN RD 桃林路

昌 邑 路

MINSHENG RD 民生路

BINJIANG AVENUE

CHANGYI ROAD

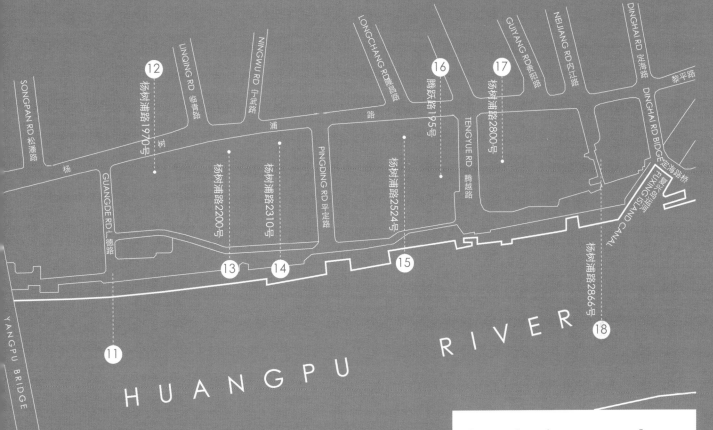

杨浦滨江历史
建筑

Historical Buildings along Yangpu Waterfront

11 / 上海电站辅机厂西厂
Shanghai Electric,Ltd.Power Station
Auxiliery Equipment Plant

12 / 上海第九棉纺织厂
Ninth Shanghai Cotton Plant

13 / 上海电站辅机厂东厂
Shanghai Electric,Ltd.Power Station
Auxiliery Equipment Plant

14 / 上海制皂厂
Shanghai Soap Factory

15 / 杨树浦煤气厂
Yangtszepoo Gas Plant

16 / 上海第十二棉纺织厂
Twelfth Shanghai Cotton Plant

17 / 杨树浦发电厂
Yangtszepoo Power Plant

18 / 上海国际时尚中心
Shanghai Fashion Centre

01

秦皇岛码头

地址：秦皇岛路 32 号
级别：杨浦区文物保护点

Qinhuangdao Wharf

Adress: No.32 Qinhuangdao Road
Grade: Cultural Heritage under the
Protection of Yangpu District

该码头原为一废旧码头，改建后专门停靠北洋航线（其中主要是上海至大连线）以及欧洲远洋航线船舶，以装卸北洋杂货、煤炭和进出口欧洲货物为主。20 世纪初，一场赴法勤工运动席卷全国。黄浦码头作为勤工俭学的主要出发地，见证了周恩来、邓小平等中共重要领导人从这里踏上寻求救国真理的道路。

The wharf was originally abandoned, and after reconstruction, it is specially used for docking ships of the Northern shipping line (mainly the Shanghai-Dalian line) and the European ocean shipping line, mainly for loading and unloading northern general cargo, coal and import and export goods of Europe. In the early 20th century, a movement to work in France swept the country. Huangpu Wharf as the main place of departure of work-study program, witnessed Zhou Enlai, Deng Xiaoping and other important leaders of the Communist Party of China embark on the road to seek the truth of national salvation from here.

02

毛麻仓库

地址：杨树浦路 468 号

Maoma Warehouse

Adress: No.468 Yangshupu Road

毛麻厂仓库始建于 1920 年，前身是德商瑞记洋行于 1895 年创办的瑞记纱厂。1918 年英商安利洋行接管经营，更名为东方纱厂。1929 年由民族资本家荣宗敬和荣德生创办的"申新纺织"收购，易名为申新七厂。中华人民共和国成立后，这里成为上海第一丝织厂，是当时全国生产真丝和人造丝的龙头企业。

Maoma Warehouse was built in 1920, whose predecessor is Arnhold Karberg & Co. Cotton Mill founded by the German business Arnhold Karberg & Co. in 1895. In 1918, British business Arnhold Brothers & Co., Ltd. took over the operation and changed its name to Dongfang Cotton Mill. In 1929, "Shenxin Textile" founded by naSung Sing Cotton Mill No.7 Plantongjing and Rong Desheng, purchased the factory and renamed it as Seventh Shenxin Factory. After the founding of the People's Republic of China, it became the First Shanghai Silk Mill and was the leading enterprise in the production of silk and rayon in China at that time.

03

上海船厂浦西分厂

地址：杨树浦路 540 号
级别：上海市优秀历史建筑

Puxi Branch of Shanghai Shipyard Co., Ltd.

Adress: No.540 Yangshupu Road
Grade: Excellent Heritage
Architecture in Shanghai

1900 年，瑞镕船厂由德商瑞记洋行创办，1912 年扩张兼并万隆铁工厂，1918 年第一次世界大战德国失败后改为英商经营，1936 年与耶松公司合并组成英联船厂股份有限公司，总厂设在瑞镕船厂，成为当时国内拥有最多船坞的修船厂。

In 1900, the New Engineering and Shipbuilding Works Ltd. was founded by German business Arnhold Karberg & Co.; in 1912, it expanded and merged with Wanlong Iron Factory; in 1918, after the Germany failed in the First World War, it was sold to British business; in 1936, it merged with Farnham & Co., S. C. to form the United Kingdom Shipyard Co., Ltd., the main plant was located in the New Engineering and Shipbuilding Works Ltd., becoming the domestic repair shipyard with the most shipyards.

04

英商怡和纱厂

地址：杨树浦路 670 号
级别：上海市优秀历史建筑

Jardine, Metheson & Co. Cotton Mill

Adress: No.670 Yangshupu Road
Grade: Excellent Heritage
Architecture in Shanghai

1896 年英商设立，是外资在沪开办的最早的工厂之一，生产的兰龙牌棉纱有一定的声誉。1911 年改属怡和纱厂股份有限公司。

Established in 1896 by British merchant, it is one of the earliest factories set up by foreign investors in Shanghai. The cotton yarn product of Lan Long brand enjoys a great popularity. In 1911, it was renamed as Jardine, Metheson Cotton Mill Co., Ltd.

05

杨树浦水厂

地址：杨树浦 830 号
级别：全国重点文物保护单位

Yangtszepoo Waterworks

Adress: No.830 Yangshupu Road
Grade: Important Monument
under the State Protection

1880 年，英商在伦敦注册成立"上海自来水股份有限公司"，并于次年在黄浦江边购地建厂，1883 年竣工。1883 年 8 月 1 日，清廷总督李鸿章开闸放水，标志着中国第一座现代化水厂正式建成，日供水量 2270 吨，可供 15 万人饮用。20 世纪 30 年代，占地面积增加了 3 倍，日供水能力达 40 万立方米，成为远东第一大水厂。

In 1880, the British registered the Shanghai Waterworks Co., Ltd. in London, and in the following year purchased land along Huangpu River to build a factory, which was completed in 1883. On 1 August 1883, Li Hongzhang, the governor of the Qing Dynasty, opened the sluice to release water, marking the official completion of China's first modern water works, with a daily water supply of 2,270 tons, drinking water for 150,000 people. In the 1930s, it increased its area by three times, and its daily water supply capacity reached 400,000 cubic meters, making it the largest waterworks in the Far East.

06

水厂栈桥

Trestle at the Waterworks

水厂栈桥长 535 米，采用钢木结构栈桥悬浮于水面上，是水厂拦污设施的一部分。栈桥以"舟桥"为设计理念，呈现船型的结构，与建成于 1883 年的英式水厂建筑隔江相望。短短的 535 米，结合江上原有的工业痕迹以及水厂的部分设施，设置了 8 个景点。

The 535-meter-long trestle is a steel and timber structure suspended above the water and is part of the pollution containment facilities of the water works. With the design concept of "boat bridge", it has a boat-like structure, facing the British waterworks building built in 1883 across the river. Within the 535 meters, there are eight tourist attractions combined with the original industrial traces on the river and part of the water works facilities.

07

杨树浦纱厂大班住宅

地址：杨树浦路 1056 号
级别：杨浦区文物保护点

Yangtszepoo Spinning Mill Residential

Adress: No.1056 Yangshupu Road
Grade: Cultural Heritage under the
Protection of Yangpu District

20 世纪初华商在此办绞花局，1915 年被英商收购，改称杨树浦纱厂。1921 年改属怡和纱厂股份有限公司，为区别于 1896 年所建的"怡和纱厂"，习称"新怡和"。新中国成立后，成为上海第一毛条厂。现尚存原大班住宅，系英式建筑，建于 1920 年前后。建筑整体造型精美，装饰、线脚细致，室内彩色玻璃、壁炉、门窗、楼梯等细部较多且保存完好。

At the beginning of the 20th century, Chinese businessmen ran a Textile Bureau here. In 1915, it was bought by British merchant and renamed as Yangshupu Cotton Mill. In 1921, it was changed to Jardine, Metheson & Co. Cotton Mill, in order to differ from the one built in 1896, also called "New Jardine, Metheson & Co.". After the founding of the People's Republic of China, it became the first wool mill in Shanghai. The existing Daban Residence is a British-style architecture, built around 1920. The overall shape of the building is exquisite with fine decoration, indoor stained glass, fireplace, doors and windows, stairs and other more details are all well preserved.

08

祥泰木行

地址：杨树浦路 1426 号
级别：杨浦区文物保护点

China Import & Export Lumber Co., Ltd.

Adress: No.1426 Yangshupu Road
Grade: Cultural Heritage under the
Protection of Yangpu District

祥泰木行，又称山打木行，由德商创立于 1884 年。其鼎盛时期，仅在上海就设有一个总栈、两个锯木厂、一个胶合板厂和一支船队，雄霸上海木材进出口市场半个世纪。

China Import & Export Lumber Co., Ltd., also known as Shanda Lumber Co., Ltd, was founded by German merchant in 1884. In its heyday, it had a master warehouse, two lumber mills, a plywood factory and a fleet in Shanghai alone, dominating the lumber import and export market of Shanghai for half a century.

09

永安栈房

地址：杨树浦 1578 号
级别：杨浦区文物保护点

Wing On Warehouse

Adress: No.1578 Yangshupu Road
Grade: Cultural Heritage under the
Protection of Yangpu District

旅澳华侨郭东、郭顺集资 3000 万元于 1921 年创建永安第一棉纺织厂。厂区建成时占地面积 4 万平方米，有 3 万纱锭、700 台织机。现存永安第一棉纺织厂仓库两座，建于 1930 年，旧称永安栈房。

Guo Dong and Guo Shun, two overseas Chinese in Australia, raised funds of 30 million yuan in 1921 to build the First Wing On Cotton Plant. When the factory was completed, it covered an area of 40,000 square meters, with 30,000 spindles and 700 looms. The existing two warehouses of First Wing On Cotton Plant are built in 1930, and used to be called Wing On Warehouse.

10

杨浦大桥

Yangpu Bridge

杨浦大桥连接杨浦区与浦东新区的过江通道，大桥于 1993 年 10 月 23 日通车运营。杨浦大桥主桥全长 1172 米，桥面为双向六车道城市快速路。杨浦大桥分别由水上主桥、陆地浦东、浦西引桥、倒 Y 形桥塔及其立交匝道组成。杨浦大桥为一跨过江的双塔双索面斜拉桥，采用钢梁与钢筋混凝土预制板相结合的叠合梁结构。为迎接中国国际进口博览会，杨浦大桥已于 2018 年 8 月完成景观提升改造。杨浦大桥拥有平时、节假日、深夜等不同模式的景观灯光，在夜间展示其矫健挺拔的身姿。

Yangpu Bridge, which connects Yangpu District with Pudong New Area across the river, was opened to traffic on 23 October 1993. The main bridge of Yangpu Bridge is 1,172 meters long. The deck of the bridge is a two-way six-lane urban expressway. Yangpu Bridge consists the main overwater bridge, land of Pudong, approach bridge of Puxi, the inverted Y-shaped bridge tower and its overpass ramp. Yangpu Bridge is a cable-stayed bridge with two towers and two cable-planes across the river, with a composite beam structure of steel beams and precast reinforced concrete slabs. In preparation for the China International Import Expo, the landscape of Yangpu Bridge was upgraded in August 2018. It now has different modes of landscape lights on weekdays, holidays and late at night, showing its vigorous and upright posture.

11

上海电站辅机厂西厂

Shanghai Electric,Ltd. Power Station Auxiliery Equipment Plant

1906 年，美商奇异电器公司及花旗银行等在美国注册成立慎昌洋行。1921 年，慎昌洋行在杨树浦路选址 22 万平方米，建成了美商慎昌洋行杨树浦工场，开始生产电风扇、电冰箱、手表等金属构件。1950 年，慎昌洋行在杨树浦路又建立西厂区，1980 年 5 月，改为上海电站辅机厂，成为我国最大的电站辅机专业设计制造厂。

In 1906, the American business General Electric and Citibank registered in the United States to establish Anderson, Meyer & Co. In 1921, it located 220,000 square meters site on Yangshupu Road to build the Yangshupu Factory of Anderson, Meyer & Co., which began to produce electric fans, refrigerators, watches and other metal components. In 1950, Anderson, Meyer & Co. set up the West Plant on Yangshupu Road. In May 1980, it was renamed Shanghai Power Station Auxiliary Machinery Plant, which has become the largest professional design factory of power station auxiliary machinery in China.

12

上海第九棉纺织厂

地址：杨树浦路 1970 号
级别：上海市优秀历史建筑

Ninth Shanghai Cotton Plant
Adress: No.1970 Yangshupu Road
Grade: Excellent Heritage Architecture in Shanghai

1895 年，官督商办大纯纱厂成立，后因经营不善，于 1905 年出租给日商三井洋行经营。1906 年，三井洋行正式收购大纯纱厂，改称三泰纱厂。1908 年三井洋行将三泰与兴泰两家纱厂合并，正式改名上海纺织株式会社，简称上海纱厂。1949 年 10 月，改组为上海第九棉纺织厂。

In 1895, Dachun Cotton Mill was established by businessman but run by the government. Later, due to poor management, it was leased to Mitsui & Co., Ltd. in 1905 for operation. In 1906, Mitsui & Co., Ltd. formally acquired the Dachun Cotton Mill and renamed it as Santai Cotton Mill. In 1908, Mitsui & Co., Ltd. merged the two mills, Santai and Xingtai, and officially renamed as Shanghai Cotton Manufacturing Co., Ltd., generally shorten to be Shanghai Cotton Mill. In October 1949, it was reorganized as Ninth Shanghai Cotton Plant.

13

上海电站辅机厂东厂

地址：杨树浦路 2200 号

Shanghai Electric,Ltd. Power Station Auxiliery Equipment Plant
Adress: No.2200 Yangshupu Road

上海电站辅机厂东厂以铜梁路为界，1949 年前，分别属于一战时期建立的三井木工厂及 1921 年间建立的慎昌洋行工厂。1952 年，两厂与慎昌洋行在杨树浦路西侧另一厂区合并为浦江机器厂，本区域称之为东厂，并于 1953 年成立上海锅炉厂。1980 年电站辅机由锅炉厂分出独立经营，分为东西两厂，成为国内规格最大、品种最多的电站辅机制造专业企业，以及我国核电设备制造的骨干企业。

East Plant of Shanghai Power Station Auxiliary Machinery Plant is bordered by Tongliang Road. Before 1949, it separately belonged to Mitsui Wood Factory established during the First World War and Anderson, Meyer & Co. Factory established in 1921. In 1952, the two factories merged with another factory of Anderson, Meyer & Co. on the west side of Yangshupu Road to form Pujiang Machinery Plant, which was called the East Plant in this area, and Shanghai Boiler Plant was established in 1953. In 1980, the Power Station Auxiliary Machinery Plant was separated from the Boiler Plant and operated independently. It was divided into two plants, the east and the west, and became the professional manufacturing enterprise of power station auxiliary machinery with the largest specifications and most varieties in China, as well as a backbone enterprise of nuclear power equipment manufacturing in China.

14

上海制皂厂

地址：杨树浦路 2310 号

Shanghai Soap Factory
Adress: No.2310 Yangshupu Road

上海制皂厂创建于 1923 年，原名为英商中国肥皂股份有限公司上海分公司。1952 年，该厂由上海市人民政府接管，改名为中国肥皂公司。1955 年，定名为上海肥皂厂。上海肥皂厂是唯一生产系列肥皂的国家一级企业。

Shanghai Soap Factory was founded in 1923 as the Shanghai Branch of the British China Soap Co., Ltd. In 1952, the factory was taken over by the Shanghai Municipal People's Government and renamed China Soap Company. In 1955, it was named as Shanghai Soap Factory. Shanghai Soap Factory is the only first-class national enterprise producing serial soap.

15

杨树浦煤气厂

地址：杨树浦路 2524 号
级别：上海市文物保护单位

Yangtszepoo Gas Plant

Adress: No.2524 Yangshupu Road
Grade: Cultural Heritage under the Protection of Shanghai

上海煤气公司杨树浦工场旧址即杨树浦煤气厂，于 1932 年筹建，1934 年 8 月投产，后成为远东第一大煤气厂，建成时日产煤气 11.3 万立方米，包括水煤气 5.67 万立方米，占全市煤气消费量的 80%。

Yangshupu Gas Plant, the former site of Shanghai Gas Company, was prepared to be build in 1932 and put into production in August 1934. It later became the largest gas plant in the Far East, with a daily output of 113,000 cubic meters of gas, including 56,700 cubic meters of water gas, accounting for 80% of the city's gas consumption.

16

上海第十二棉纺织厂

地址：腾跃路 195 号

Twelfth Shanghai Cotton Plant

Adress: No.195 Tengyue Road

十二棉前身为日本纺绩株式会社所属的大康纱厂，第一纱厂于 1920 年筹建，1921 年建成。上海解放后，由市军管会接管，1950 年 7 月 1 日更名为国营上海第十二棉纺织厂。上海第十二棉纺织厂至 1998 年关闭，历经 78 年历史。

Twelfth Shanghai Cotton Plant was formerly known as Dakang Textile Mill, which was affiliated to Nippon Cotton Spinning Co., Ltd. The first cotton mill was prepared in 1920 and completed in 1921. After the liberation of Shanghai, it was taken over by the Municipal Military Control Commission, and on 1 July 1950, it was renamed the State-run Twelfth Shanghai Cotton Plant. It has a history of 78 years until its closing in 1998.

17

杨树浦电厂

地址：杨树浦路 2800 号
级别：上海市文物保护单位

Yangtszepoo Power Plant

Adress: No.2800 Yangshupu Road
Grade: Cultural Heritage under the Protection of Shanghai

上海工部局电气处新厂旧址即杨树浦发电厂，筹建于 1910 年，1913 年 4 月 12 日正式发电，是中国建造较早的大型火电厂。初时装机容量为 10400 千瓦，到 1924 年，装机容量达 12.1 万千瓦，成为当时远东第一大电厂。2010 年发电厂正式关停，随着 2015 年 6 月杨浦滨江南段公共空间改造工程在电厂段的开始，原有的净水池装置被改造成为咖啡厅；而电厂段最东端的码头上，原本的干灰储煤灰罐则被改造为灰仓美术馆。

Yangshupu Power Plant was the former site for the new plant of Electrical Department of Shanghai Ministry of Industry, which was prepared to build in 1910 and started operation on 12 April 1913. It is one of the earliest large-scale thermal power plants built in China. At the beginning, the installed capacity was 10,400 kilowatts; by 1924, the installed capacity reached 121,000 kilowatts, making it the largest plant in the Far East at that time. In 2010, the power plant was officially shut down, With the beginning of the public space reconstruction project in the south section of Yangpu Waterfront in June 2015, the original device of water purification tank was transformed into a café, while on the wharf at the easternmost end of the power plant section, the original dry ash storage tank has been transformed into the Ash Gallery.

18

上海国际时尚中心

地址：杨树浦路 2866 号
级别：上海市文物保护单位

Shanghai Fashion Centre

Adress: No.2866 Yangshupu Road
Grade: Cultural Heritage under the Protection of Shanghai

日本大坂东洋纺织株式会社的上海工场，于 1921 年建造，1936 年改名为裕丰株式会社，以生产著名的"龙头细布"为主。1946 年改名中国纺织建设公司第十七棉纺织厂。1949 年改名上海第十七棉纺织厂。是全国第一家批量生产棉型腈纶针织纱的企业。

The Shanghai workshop of Osaka Toyobo Textile Co., Ltd. was built in 1921 and renamed as Yu Fong Co., Ltd. in 1936. It mainly produces the famous "Dragon Head Fine Cloth". In 1946, it was renamed as Seventeenth Cotton Plant of China National Textile Company of Construction. In 1949, it was renamed Seventeenth Shanghai Cotton Plant. It is the first enterprise that produces acrylic fiber knitting yarn of cotton type in China.

SUSAS 学院
SUSAS College

为进一步加强上海城市空间艺术季的行业引领，为专业技术人员提供学术交流和互动的平台；积极发挥公众教育作用，吸引市民更多地参与空间艺术季，走进城市空间，讨论城市议题，以主人翁的精神，参与城市建设和城市治理，上海城市空间艺术季自第二届起设立公众活动（SUSAS 学院）版块，并不断努力打造具有特色和吸引力，有辨识度和价值导向的品牌活动。SUSAS 学院不是传统意义上的有物理空间的教育场所，而是有关城市、公共空间及公共艺术的，面向大众且开放的教育资源，是协作学习及互动的平台。

2019 空间艺术季的公众活动（SUSAS 学院）版块围绕"滨水空间给人类带来美好生活"的主题方向，通过学术交流、市民宣传、儿童教育及其他活动，让不同专业领域、不同年龄层次的人在主展场空间"相遇"，讨论"滨水空间"的更新，体验"城市艺术"的魅力，感知"杨浦滨江展区"的历史，展开对"未来生活"的畅想。

2019 空间艺术季自 2019 年 9 月 29 日开幕到 2019 年 12 月 17 日闭幕，在面向公众展览的 80 天里，公众活动版块共组织 144 场活动，参与人数共 1 万余人次，平均每天约举办 1.75 场活动，平均每场活动约 70 多人次。其中，学术交流活动 20 场，如大师讲坛、同济设计周系列讲座、设计教育会议、H+A 华建筑学术年会等；市民宣传活动 10 场，如公共艺术作品公众导览、滨江空间景观公众导览、社区规划师沙龙、杨浦滨江摄影大赛与优秀作品展、城市空间行走体验、艺术护照打卡等；儿童教育活动 78 场，如小小导赏员、小小观察员、小小建筑师、小小规划师、儿童创意工作坊、乐高积木搭建等；其他类型活动 35 场，如绿之丘临展、大学生创意市集、种子魔方活动、滨江定向赛等。

144 场活动分别举办于杨浦滨江的室内外不同类型的空间场所中，如主展馆片区毛麻仓库、小白楼、1 号船坞、2 号船坞，杨浦滨江特色空间中的绿之丘、渔人码头、共生构架、皂梦空间、边园、杨树浦电厂、国际时尚中心等。

In order to strengthen the industry leadership of Shanghai Urban Space Art Season, provide professional personnel with a platform for academic exchanges and interaction. Actively playing the role of public education and attracting citizens to participate more in the SUSAS, get in touch with the urban space, and discuss urban issues. Leading the public to take a role and responsibility as a citizen to participate in urban construction and urban governance. Shanghai Urban Space Art Season has set up a public interactive (SUSAS College) section in 2017 and strives to create distinction and attractivity constantly, take place outstanding and value-oriented brand activities. SUSAS College is not a conventional place with physical space in the traditional sense, but an open resource, connecting with the city, public space, and public art, to the ordinary people. It is a platform for collaborative learning and interaction.

The public activity (SUSAS College) section of 2019 SUSAS focused on the theme of "waterfront space brings a better life to mankind", through academic exchanges, public publicity, children's education, and other activities to enable people of different professional fields and ages to "encounter" in the main exhibition space to discuss the renewal of "waterfront space", experience the charm of "urban art", perceive the history of "Yangpu Waterfront exhibition area" and imagine "future life".

2019SUSAS opened on September 29, 2019, and closed on December 17, 2019. In the 80 days of public exhibition, 144 events were organized in the public activities section, with more than 10000 visits. On average, about 1.75 events will be held every day and more than 70 people participated in each event. Among them, there were 20 academic exchange activities, such as master forum, Tongji design-week series lectures, design education conference, H + A Hua Architecture academic annual meeting, etc; 10 public publicity activities were held, such as public guide to public artworks, public guide to waterfront space landscape, community planner salon, Yangpu waterfront photography competition and excellent works exhibition, urban spacewalking, art passport tagging, etc; 78 children's educational activities were held, such as little tour guides, little observers, little architects, little planners, children's creative workshops, Lego construction, etc; 35 other types of activities were held, such as The Green Hill temporary exhibition, college students' innovative market, seed cube activity, waterfront orienteering, etc.

144 events were held in different types of indoor and outdoor space in Yangpu Waterfront, such as Maoma warehouse, Xiaobai building, No. 1 shipyard and No. 2 shipyard in the main exhibition hall area; The Green Hill, Fisherman's Wharf, Pavilion of Symbiosis, Fantasy Bubble, Riverside Passage, Yangtszepoo Power Plant, Shanghai Fashion Center, etc.

毛麻仓库

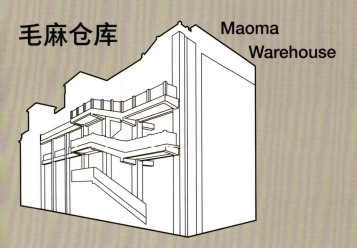

Maoma Warehouse

学术交流类活动一览

名称：建筑规划版块开幕论坛
时间：2019/9/29

名称：大师讲坛第一场——看见：
这座城市的温度 纪录片专场
时间：2019/11/5

名称：大师讲坛第三场——撞见：
公共艺术，换个角度看世界
活动时间：2019/11/24

儿童教育类活动一览

名称：儿童艺术工作坊
时间：2019/11/24

名称：小小建筑师
时间：2019/10/19

讲座：力之力：上海杨树浦发电厂的感知与设计
时间：2019/11/22

讲座：苏州河的故事
时间：2019/11/10

名称：趣城课堂
时间：2019/11/30

名称：新书发布会
时间：2019/11/23

名称：第二届中欧绘本与创新论坛 & 文化创意
产业与创新教育研究国际论坛
时间：2019/10/13

讲座：杨浦区社区规划师培训
时间：2019/11/17

名称：《H+A 华建筑》杂志年度工作会议
时间：2019/11/26

名称：《H+A 华建筑》学术年会——华山论建
大师讲堂
时间：2019/12/20

讲座：公共性与剩余空间
时间：2019/11/23

第二届中欧绘本与创新论坛 & 文化创意产业与创新教育研究国际论坛

小小导览员活动现场

插画展儿童工作坊

大师讲坛活动现场

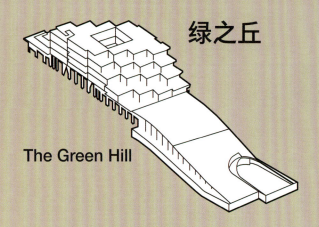

绿之丘

The Green Hill

名称：声音造物展
时间：2019/10/14—11/3

名称：中非设计智慧展
时间：2019/10/14—11/3

名称：Ken Friendman：92 件事
时间：2019/10/14—11/3

名称："连接感" 空间设计展
时间：2019/10/14—11/3

名称：Closing the loop 循环设计展
时间：2019/10/14—11/3

名称：展陈的力量
时间：2019/10/14—11/3

名称：作为经验的流动性
时间：2019/10/14—11/3

名称：第二届中欧当代插图与影像展
时间：2019/10/12—11/12

名称：杨浦七梦
时间：2019/09/28—11/30

名称：比特纪：空间链、公共性欲技术批判
时间：2019/9/29—10/28

名称：合成空间与感官
时间：2019/10/5

名称：SUSAS2019 摄影展：对话
时间：2019/12/13—2020/1/19

名称："聚场"——上海大学上海美术学院设计系课程教学展
时间：2019/11/20—11/28

名称：文创实验室：FABZIGN LAB
时间：2019/9/29—11/30

名称：字眼展
时间：2019/9/29—11/30

名称：地缘相遇：一带一路的景观图绘
时间：2019/10/3—11/30

名称：涟漪家园：百名青年家园水故事展
时间：2019/11/10—11/30

第二届中欧当代插图与影像展现场

展陈的力量展览现场照

声音造物展展览现场照

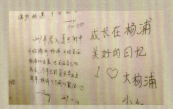

杨浦七梦展览现场

338

共生构架

Symbiosis Architecture

名称：乐高积木搭建
时间：2019/10/26—10/27

"共生·见面"乐高积木搭建活动现场

边园
Riverside
Passage

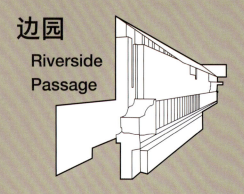

名称：种子魔方活动
时间：2019/11/10

"种子大魔方"活动现场

船坞

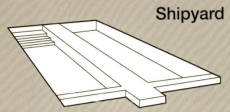

Shipyard

2019/09/29-11/30 上海/杨浦滨江 Shanghai Yangpu Riverside

相遇 encounter

2019 上海城市空间艺术季 SHANGHAI URBAN SPACE ART SEASON

总策展人/北川富朗 ARTISTIC DIRECTOR / FRAM KITAGAWA 规划建筑策展人/阮昕 空间艺术策展人/川添善行 总建筑师/章明

2019 上海城市空间艺术季开幕式
活动时间：2019/9/28

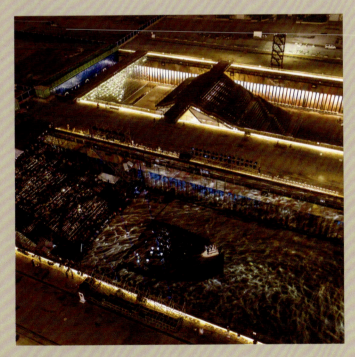

2019 上海城市空间艺术季开幕式现场

杨浦滨江沿线

Yangpu Waterfront

2019"汇创滨江"国际大学生创意市集

名称: 2019"汇创滨江"国际大学生创意市集
暨文化创意产业与创新教育研究国际论坛
时间: 2019/10/13

名称: 2019 上海街艺节开幕系列杨浦专场
时间: 2019/10/13

名称: WOD LIVE 2019 世界舞蹈大赛巡回演出
时间: 2019/10/19-10/20

名称: 空间艺术快闪导览
时间: 2019/11/17

名称: 空间艺术季定向赛
时间: 2019/11/30

名称:《城市的野生》工作坊
时间: 2019/8/24—8/25

2019"汇创滨江"国际大学生创意市集讲座

街艺节现场

主视觉设计
Main Visual Design

作品点位

+

铜　+　水　+　铁　+　沙

设计者
Designer

韩家英
Han Jiaying

1961 年出生于中国天津
1986 年毕业于西安美术学院
1993 年创立韩家英设计公司
国际平面设计联盟（AGI）会员
中央美术学院城市设计学院客座教授
2003 年在法国举办《天涯》专题设计个展
2012 年首届中国设计大展平面策展人
2012-2014 年在中央美院美术馆、深圳华美术馆、上海洛克外滩源举办"镜像·韩家英设计展"
2016 年受邀参加华沙海报双年展五十周年特别展"50/50/50"
曾荣获亚洲最具影响力设计大奖金奖
福布斯 2015 中国最具影响力的设计师
多项作品收藏于包括 V&A 博物馆在内的英国、法国、德国、丹麦、日本、中国等国际艺术机构

Born in Tianjin in 1961
Graduated from Xi'an Academy of Fine Arts in 1986
Established Han Jiaying Design Company in 1993
Member of the Alliance Graphique Internationale (AGI)
Visiting Professor, School of Urban Design, Central Academy of Fine Arts
Held a solo exhibition of "The End of the World" in France in 2003
Plan Curator of the first China Design Exhibition 2012
2012-2014 held the "Mirror · Han Jiaying Design Exhibition" at the CAFA Art Museum, OCT Art & Design Gallery, and Shanghai ROCKBUND
In 2016, he was invited to participate in the special exhibition ""50/50/50"" of the 50th Anniversary of the Warsaw Poster Biennale
Won the DFA Design for Asia Awards
Forbes 2015 China's most influential designer
Works are collected by many international art institutions in the United Kingdom, France, Germany, Denmark, Japan, China, including V&A Museum

艺术 / 相遇滨江

滨江岸线｜二十多位艺术家｜数十个点位

圆点——点位、源点、坐标、人物

以艺术碰撞、多元融合、亮色汇聚的方式合并成一种新的形态

进一步把基于滨江、更走向世界的艺术想像无限延伸

呈现一种滨江之上，普遍认知以外的超艺术化感知与城市化联动

Art/Encounter at the Waterfront

Waterfront Shoreline | More than 20 Artists | More than 10 Sites

Circle — Site, Origin, Coordinate, Figure

Through artistic jamming, diverse fusion, highlight converging to merge into a new form

To further extend the artistic imagination based on the waterfront towards the world

It presents a linkage between the ultra-artistic perception and urbanization beyond the general cognition on the waterfront

花絮
Titbits

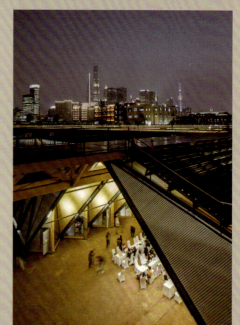

摄影者索引
Photographers Index

其余作品过程图和照片均由参展人提供

The other process images are all supplied by exhibitors.

致谢
Acknowledgements

两年一届的空间艺术季是艺术介入城市，激活地区更新，进而推动城市空间提升品质的实践行动。本届空间艺术季聚焦"滨水空间为人类带来美好生活"这一世界性话题，努力推动展览从室内走向室外开放空间；从作品借展转向作品定制并永久留存一批作品；同时融合所有空间设计力量，全力呈现杨浦滨江公共空间这一底蕴深厚且富有魅力的空间作品。

2018 年 8 月，上海城市公共空间设计促进中心经过前期调研，提出了 2019 空间艺术季选址杨浦滨江的方案，得到了杨浦区政府的积极响应；2018 年 9 月，空间艺术季与杨浦区政府达成了选址于杨浦滨江 5.5 公里滨江空间的合作意向；2018 年 9 月 24 日，北川先生第一次来到杨浦滨江现场踏勘，即被滨江公共空间深深吸引了。在以后的近一年的工作期间，总策展人北川富朗先生经常感叹每次走现场都不一样，这里有房子拆掉了，那里有景观建好了，滨江面貌都日新月异。开展的两个多月以来，艺术季主展场 5 万多名参观者，滨江 5.5 公里公共空间吸引 32 万市民游客前来参观，是艺术季得到公众认可的最好例证。

本届艺术季在杨浦滨江的成功举办，为文化引领的发展模式积累了宝贵的经验，永久艺术作品成为滨江文化新地标，也拓展了滨江发展的文化平台，必将对滨江发展带来深远影响和长久动力。这其中饱含着各界人士的努力付出。总策展团队和总建筑师团队承担了本次艺术季的核心工作，为此次展览邀请参展人和参展作品，改造场地和展馆，体现了高超的专业能力。杨浦滨江的高品质设计得益于这些高水平专业设计单位的倾力付出，大到整体有序的工业和生态相容的景观风貌，小到高桩码头上的大树、漫步道边的垃圾桶，都令人印象深刻。还有默默无闻的执行团队、志愿者团队，他们在幕后的艰辛工作，得以保障展览的顺利开幕与运维。此外，还要感谢持续关注艺术季的媒体朋友，是他们对艺术季的宣传报道让艺术季的光彩被更多的人了解。

未来的日子，我们希望与各界人士继续携手共进，共同发展这一城市品牌活动，为弘扬上海城市文化，打造具有世界影响力的国际化大都市增光添彩！

The biennial space art season is a practical action for art to intervene in the city, activate the regional renewal, and then improve the quality of urban space. This space art season focuses on the global topic of "how waterfronts bring wonderful life to people", and strives to promote the exhibition from indoor to outdoor open space; from loaning works for exhibition to customization and permanent retention of a number of works; and at the same time, integrating all the space design forces to present the profound and charming space work of Yangpu Waterfront Public Space.

In August 2018, after preliminary research, Shanghai Design & Promotion Centre for Urban Public Space put forward the plan of choosing Yangpu Waterfront as the site for 2019 Space Art Season, which received positive response from the Yangpu District Government. In September 2018, Space Art Season reached a cooperation intention with the government of Yangpu District to locate the event at a 5.5-kilometer-long waterfront space in Yangpu. On 24 September 2018, the first timr Mr. Fram Kitagawa came to Yangpu Waterfront and was deeply attracted by the public space along the riverside. During his nearly one year's work here, Mr. Fram Kitagawa, the chief curator of the exhibition, often raved that the site was different every time he visited: some buildings were demolished here, some landscapes were built there, and the riverside was changing with every passing day. Over the past two months since its launch, more than 50,000 visitors have visited the main exhibition site of the art season, and 320,000 citizens have visited the 5.5 km public space along the riverside, which is the best prove of the art season being recognized by the public.

The successful holding of this art season in Yangpu Waterfront has accumulated valuable experience for the culture-led development model. The permanent art works have become the new landmark of waterfront culture and expanded the cultural platform for the development of the waterfront, which will surely bring far-reaching influence and long-term impetus to the development of this area. This is full of the efforts of people from all walks of life. The chief curator team and the chief architect team undertook the core work of the art season, inviting exhibitors and artworks to participate in the exhibition, transforming the site and exhibition hall, reflecting their superb professional ability. The high-quality design of Yangpu Waterfront is benefited from the efforts of these high-level professional design institutions, from the overall orderly industrial and ecological landscape, to the small trees on the piled wharf and the trash cans by the promenade, all of them are impressive. And the unknown executive team, volunteer team, their hard work behind the scenes ensures the smooth opening and operation of the exhibition. In addition, we also have to thank the press who continue to pay attention to the art season. It is their publications and reports on the event that make the splendor of the art season known to more people.

In the future, we look forward to continuously work together with people from all walks of life to jointly develop this city brand activity, to promote the city culture of Shanghai, to build an international metropolis with great world！

鸣谢单位

2019 上海城市空间艺术季工作团队

工作小组
上海市规划和自然资源局风貌管理处
上海市规划和自然资源局财务与资金管理处
上海市规划和自然资源局公众参与处
上海市文化和旅游局艺术处
上海城市公共空间设计促进中心

杨浦区政府办公室
杨浦区财政局
上海市黄浦江杨浦段滨江综合开发指挥部办公室
杨浦区规划和自然资源局
杨浦区文化和旅游局
上海杨浦滨江投资开发有限公司

总建筑师团队
原作设计工作室
致正建筑工作室
刘宇扬建筑事务所
大舍建筑设计事务所
同济大学建筑设计研究院（集团）有限公司都境建筑设计院
上海优德达城市设计咨询有限公司
大观景观设计

永久艺术作品制作协调团队
欣稚锋艺术发展（上海）有限公司

主视觉设计团队
韩家英设计有限公司

布展执行团队
上海风语筑展示股份有限公司

运营团队
上海木宁芙文化发展有限公司

视频拍摄制作团队
ACTION MEDIA

主题曲创作团队
上海彩虹室内合唱团

案例展主承办单位

浦东新区实践案例展
主办单位：上海市浦东新区人民政府
承办单位：浦东新区规划和自然资源局
策展人：奚文沁、卞硕尉、杨帆

徐汇区上海西岸实践案例展
主办单位：上海市徐汇区人民政府
承办单位：徐汇区规划和自然资源局、徐汇区文化和旅游局、西岸集团
策展团队：西岸文化艺术季

闵行区浦江第一湾公园实践案例展
主办单位：上海紫竹高新技术产业开发区、上海市闵行区吴泾镇人民政府
承办单位：上海交通大学设计学院、Let's Talk 学术论坛
策展人：张海翱、戴春

普陀区 M50 创意园实践案例展
主办单位：上海市普陀区人民政府
承办单位：普陀区规划和自然管理局、普陀区长寿路街道办事处
策展团队：上海交通大学城市更新保护创新国际研究中心、上海安墨吉建筑规划设计有限公司

长宁区苏州河实践案例展
主办单位：上海市长宁区人民政府
承办单位：长宁区规划和自然资源局、长宁区建设和管理委员会、长宁区绿化市容局、周家桥街道
策展人：李丹锋、周渐佳

静安区彭越浦河岸景观改造实践案例展
主办单位：上海市静安区人民政府
承办单位：静安区规划和自然资源局、静安区彭浦镇人民政府
策展人：金江波、张承龙

嘉定区实践案例展
主办单位：上海市嘉定区人民政府
承办单位：嘉定区规划和自然资源局
策展人：周芳珍

青浦环城水系公园空间艺术实践案例展
主办单位：上海市青浦区人民政府
承办单位：上海淀山湖新城发展有限公司
策展人：施皓

虹口区实践案例展
主办单位：上海市虹口区人民政府
承办单位：虹口区规划和自然资源局、虹口区文化和旅游局、虹口区北外滩街道
策展团队：香港大学上海学习中心

松江新城实践案例展
主办单位：上海市松江区人民政府
承办单位：松江区规划和自然资源局、松江区文化和旅游局、松江区体育局、松江新城建设发展有限公司
策展人：翟伟琴

奉贤新城实践案例展
主办单位：上海市奉贤区人民政府
承办单位：奉贤区规划和自然资源局
策展人：冯路

金山区漕泾镇水库村实践案例展
主办单位：上海市金山区人民政府
承办单位：漕泾镇人民政府
策展人：苏冰、董楠楠

联合展主办单位

"相遇·贵州路"联合展
主办单位：黄浦区南京东路街道、百联集团时尚中心、新光影艺苑、上海金外滩(集团)发展有限公司、上海大光明文化(集团)有限公司、上海市商贸旅游学校

"花开上海"联合展
主办单位：上海四叶草堂青少年自然体验服务中心

"陆家嘴滨江金融 HARBOUR CITY 公共艺术景观装置展"联合展
主办单位：中船置业有限公司 上海瑞明置业有限公司

"第六届全国大学生公共视觉优秀作品双年展"联合展
主办单位：上海市规划和自然资源局、普陀区人民政府
承办单位：普陀区规划和自然资源局

"隽永墨韵大华银行水墨艺术展"联合展
主办单位：大华银行（中国）有限公司

"2019 多瑙河对话艺术节梦之声—丝路上的中国当代艺术"联合展
主办单位：上海虹庙艺术中心

"渡·爱"外滩艺术计划联合展
主办单位：上海贝思诺广告有限公司

"合成空间＆感官"联合展
主办单位：英国哈德斯菲尔德建筑学院

"盲点艺术展"联合展
主办单位：阆风艺术

"步履不停：1995—2019 中国当代艺术的城市叙事"联合展
主办单位：上海多伦现代美术馆

"一生万物要有光艺术展"联合展
主办单位：维亚景观、那行零度空间

"生境花园：给城市野生动植物建个家"联合展
主办单位：大自然保护协会、桃花源生态保护基金会

"符号上海 相遇·水岸"联合展
主办单位：上海城市规划展示馆、上海市青少年活动中心

鸣谢单位及个人

上海国际艺术节中心
上海彩虹室内合唱团
华东建筑集团股份有限公司
光明食品（集团）有限公司
澎湃新闻市政厅、第六声
同济大学建筑设计研究院（集团）有限公司
上海那行文化传媒有限公司
Playable Design
锐字家族
上海瞳初文化传播有限公司
上海彩虹青少年发展中心
杨浦邮政分公司民星邮政支局
上海方驰建设有限公司
上海杨树浦智慧物业管理有限公司
上海杨树浦文化创意产业有限公司
上海杨树浦城市开发建设有限公司
东方渔人码头
葛珺
吴巧云

Acknowledgements for Units and Individuals

2019 SUSAS Teams

· Working Group ·
Landscape Division, Shanghai Urban Planning and Natural Resources Bureau
Finance and fund management division, Shanghai Urban Planning and Natural Resources Bureau
Public participation Division, Shanghai Urban Planning and Natural Resources Bureau
Art Division, Shanghai Municipal Administration of Culture and Tourism
Shanghai Design & Promotion Centre for Urban Public Space

Yangpu District People's Government Office
Yangpu District Finance Bureau
Shanghai Huangpu River Yangpu Section Riverside Comprehensive Development headquarters Office
Yangpu District Planning and Natural Resources Bureau
Yangpu District Municipal Administration of Culture and Tourism
Shanghai Yangpu Waterfront Investment and Development Ltd.

· Chief Architect Team ·
Original Design Studio
Atelier Z+
Atelier Liu Yuyang Architects
Atelier Deshaus
Dujing Architectural Design Institute ,Tongji Architectural Design（Group）Co.,Ltd.
UrbanDATA
Da Landscape

· Survived Art Works of Production Coordination Team ·
Art Pioneer Studio

· Main Visual Design Team ·
Han Jiaying Design Co., Ltd.

· Exhibition Executive Team ·
Shanghai fengyuzhu Exhibition Co.,Ltd.

· Operation Team ·
Shanghai muningfu Culture Development Co., Ltd.

· Video Production Team ·
ACTION MEDIA

· Theme Song Composing Team ·
Rainbow Chamber Singers

Site Projects Host Unit

· Site Project at Pudong New Area ·
Host: People's Government of Pudong New Area, Shanghai
Organizer: Pudong New Area Planning and Natural Resources Bureau
Curators: Xi Wenqin, Bian Shuowei, Yang Fan

· Site Project at West Bund, Xuhui District ·
Host: People's Government of Xuhui District, Shanghai
Organizers: Xuhui District Urban Planning and Natural Resources Bureau, Xuhui District Culture and Tourism Bureau, Shanghai West Bund Development (Group) Co., Ltd
Curatorial Team: Art' West Bund

· Site Project at Pujiang First Bay Park, Minhang District ·
Host: Zizhu National Hi-tech Industrial Development Park, People's Government of Wujing Town, Minhang District, Shanghai
Organizers: School of Design,Shanghai Jiao Tong University , Let's Talk
Curators: Zhang Hai'ao, Dai Chun

· Site Project at M50 Creative Park, Putuo District ·
Host: People's Government of Putuo District, Shanghai
Organizer: Putuo District Planning and Natural Resources Bureau, Changshou Road Sub-district Office
Curatorial Teams: Shanghai Jiaotong University International Research Centre for Creative Urban Regeneration and Protection, AMJ Shanghai

· Site Project at Suzhou Creek Waterfront, Changning District ·
Host: Changning Distric People's Government of Changning District, Shanghai
Organizers: Planning and Natural Resources Administration Bureau of Changning District, Construction and Traffic Committee of Changning District, Administration of Greening and City Appearance of Changning District, Zhoujiaqiao Sub-district Office
Curators: Li Danfeng, Zhou Jianjia

· Site Project in Jing'an District: Landscape Renovation of Pengyuepu Waterfront ·
Host: People's Government of Jing'an District, Shanghai
Organizers: Jing'an District Planning and Natural Resources Bureau,Pengpu Town Government of Jing'an District
Curators: Jin Jiangbo, Zhang Chenglong

· Site Project in Jiading District ·
Host: People's Government of Jiading District, Shanghai
Organizer: Jiading District Planning and Natural Resources Bureau
Curator: Zhou Fangzhen

· Site Project in Qingpu District: Space Art Practice of Qingpu Round-city Water System Park ·
Host: People's Government of Qingpu District, Shanghai
Organizer: Shanghai Lake Dianshan Newtown Development Co., Ltd.
Curator: Shi Hao

· Site Project in Hongkou District ·
Host: People's Government of Hongkou District, Shanghai
Organizers: Hongkou District Planning and Natural Resource Bureau, Hongkou District Culture and Tourism Bureau, Hongkou District North Bund Sub-District Office, Shanghai
Curatorial Teams: Shanghai Study Centre, The University of Hong Kong

· Site Project in Songjiang New City ·
Host: People's Government of Songjiang District, Shanghai
Organizers: Songjiang District Planning and Natural Resources Bureau, Songjiang Administration of Culture and Tourism, Songjiang Administration of Sports, Songjiang New City Development Company
Curator: Zhai Weiqin

· Site Project in Fengxian New City ·
Host: People's Government of Fengxian District, Shanghai
Organizer: Fengxian District Planning and Natural Resources Bureau
Curator: Feng Lu

· Site Project in Shuiku Village, Caojing Town, Jinshan District ·
Host: People's Government of Jinshan District, Shanghai
Organizer: People's Government of Caojing Town
Curators: Su Bing, Dong Nannan

Joint Exhibitions Host Unit

· Encounter Guizhou Road ·
Host: Subdistrict office of East Nanjing Road, Bailian Group Fashion Centre, Xin Guang Film Art Centre, Gold Bund Group, Shanghai Grand Cinema Group, Shanghai Business & Tourism School

· Blooming Shanghai — Community Garden Plan 2040 ·
Host: Clover Nature School

· Public Art for Lujiazui HARBOR CITY ·
Host: China Shipbuilding Properties Limited, Shanghai Ruiming Real Estate Co., Ltd

· The 6th National Art Institutes And Academics Biennale for Outstanding Public Visual Artwork of Graduates and Undergraduates
Host: Shanghai Urban Planning and Natural Resources Bureau, People's Government of Putuo District

Organizer: Putuo District Urban Planning and Natural Resources Bureau

· UOB Art in Ink Exhibition ·
Host: UOB (China)

· 2019 Danube Dialogues Dream Sound — Chinese Contemporary Art on The Silk Road ·
Host: Shanghai Hong Miao Art Centre

· The Bund Art Project "Ferry Love" 2019 ·
Host: Bestknown Advertising Agency

· Synthetic Spaces & Sensorium ·
Host: Faculty of Art, Design and Architecture at University Huddersfield

· Blind Spot ·
Host: LEVANT ART

· A Turning Moment / City Narrative of Chinese Contemporary Art from 1995-2019 ·
Host: Shanghai Duolun Museum of Modern Art

· Habitat Garden: Let's Give Wildlife a Home in Our City Host ·
Host: The Nature Conservancy, Paradise International Foundation

· LIGHTENING the Living Future ·
Host: VIASCAPE design, Zero Degree Play Ground

· Iconic Shanghai Encounter & Waterfront ·
Host: Shanghai Urban Planning Exhibition Centre, Shanghai Activity Centre for Youngsters

Acknowledgement for Units and Individuals

China Shanghai international arts festival
Rainbow Chamber Singers
Arcplus Group PLC
Bright Food (Group) Co., Ltd.
City Hall, Sixth Tone, The Paper
Tongji Architectural Design(Group) Co.,Ltd.
Shanghai Nextmixing Cultural Media Co., Ltd.
Playable Design
REEJI
Shanghai Tongchu Cultural Co., Ltd.
Shanghai Rainbow Youth Development Centre
Yangpu Post Branch Minxing Post Branch
Shanghai Fangchi Construction Co., Ltd.
Shanghai Yangshupu Wisdom property Management Co., Ltd.
Shanghai Yangshupu Cultural and Creative Industry Co., Ltd.
Shanghai Yangshupu Urban Development and Construction Co., Ltd.
Oriental Fisherman's Wharf
Ge Jun
Wu Qiaoyun

图书在版编目（CIP）数据

相遇 . 1, 2019 上海城市空间艺术季主展览 /《相遇》

编委会编 . -- 上海：东华大学出版社 , 2021

　　ISBN 978-7-5669-1962-5

　　Ⅰ . ①相⋯ Ⅱ . ①相⋯ Ⅲ . ①城市空间 - 公共空间 -

建筑设计 - 作品集 - 中国 - 现代 Ⅳ . ① TU984.11

中国版本图书馆 CIP 数据核字 (2021) 第 175000 号

相遇

2019 上海城市空间艺术季主展览

《相遇》编委会 编

策　　划：上海城市公共空间设计促进中心

　　　　　群岛 ARCHIPELAGO

特约编辑：辛梦瑶

责任编辑：高路路

设计排版：李高

版　　次：2021 年 1 月第 1 版

印　　次：2021 年 1 月第 1 次印刷

印　　刷：上海盛通时代印刷有限公司

开　　本：889mm×1194mm 1/16

印　　张：22.5

字　　数：548 000

书　　号：ISBN 978-7-5669-1962-5

定价 (两册)：418.00 元

出版发行：东华大学出版社

地　　址：上海市延安西路 1882 号

邮政编码：200051

出版社网址：http://dhupress.dhu.edu.cn

天猫旗舰店：http://dhdx.tmall.com

营销中心：021-62193056 62373056 62379558

本书若有印装质量问题，请向本社发行部调换。

Encounter

2019 Shanghai Urban Space Art Season Main Exhibition

Edited by: *Encounter* Editorial Board

ISBN 978-7-5669-1962-5

Initiated by: Shanghai Design & Promotion Centre for Urban Public Space

　　　　ARCHIPELAGO

Contributing Editor: XIN Mengyao

Editor: GAO Lulu

Graphic Design: LI Gao

Published in January 2021, by Donghua University Press,

1882, West Yan'an Road, Shanghai, China, 200051.

dhupress.dhu.edu.cn

Contact us: 021-62193056 62373056 62379558